Climate Change and European Security

It is now commonly asserted that climate change will fundamentally change international relations. It has been predicted that global warming will increase conflict within and between states, intensify food insecurity, menace the global trading system and unleash waves of migration. As a result governments are beginning to incorporate these warnings into their foreign policy initiatives. The appropriateness of their incipient responses needs to be examined in finer detail.

This book looks at the impact of climate change on European Union (EU) security policy. It explores how governments are reconfiguring their geo-strategy and broader international relations in the wake of climate change warnings. The book demonstrates that although many aspects of EU foreign policies have begun to change, 'climate security' is not yet accorded unequivocal or sufficient priority. In doing so, Youngs argues that if climate change policies are to have significant effect they can no longer be treated as a separate area of policy but must be incorporated into the more mainstream debates pertinent to EU common foreign and security policy.

This book will be of key interest to students, scholars and practitioners of climate change and policy, energy and environmental policy, EU governance and foreign policy, European studies, international relations, geography, security studies/policy and environmental economics.

Richard Youngs is Senior Associate, Carnegie Europe and Professor of International Relations at the University of Warwick, UK.

CLIMATE CHANGE AND EUROPEAN SECURITY

Richard Youngs

Routledge
Taylor & Francis Group
LONDON AND NEW YORK

First published 2015
by Routledge

4 Park Square, Milton Park, Abingdon, Oxon OX14 4RN
605 Third Avenue, New York, NY 10017

Routledge is an imprint of the Taylor & Francis Group, an informa business

British Library Cataloguing in Publication Data
A catalogue record for this book is available from the British Library

Library of Congress Cataloging-in-Publication Data
Youngs, Richard, 1968-
Climate change and European security / Richard Youngs.
pages cm
1. National security–European Union countries. 2. European Union countries–Foreign relations. 3. Common Security and Defence Policy. 4. Climatic changes–Political aspects–European Union countries. 5. Global warming–Political aspects–European Union countries. I. Title.
JZ1570.Y66 2014
355'.03304–dc23
2014010321

ISBN: 978-1-138-79727-7 (hbk)
ISBN: 978-1-138-79728-4 (pbk)
ISBN: 978-1-315-75725-4 (ebk)

Typeset in Bembo
by Cenveo Publisher Services

Contents

Abbreviations

CCS	Carbon Capture and Storage
CDM	Clean Development Mechanism
CFSP	Common Foreign and Security Policy
CIA	Central Intelligence Agency
CSDP	Common Security and Defence Policy
DDR	Disarmament, Demobilisation and Reintegration
DECC	Department for Energy and Climate Change
DfID	Department for International Development
EAS	External Action Service
EIB	European Investment Bank
EPLO	European Peacebuilding Liaison Office
ETS	Emissions Trading Scheme
EU	European Union
GDN	Green Diplomacy Network
GEF	Global Environmental Facility
GMES	Global Monitoring for Environment and Security
IEA	International Energy Agency
IES	Institute for Environmental Security
IfS	Instrument for Stability
IPCC	Intergovernmental Panel on Climate Change
IPR	Intellectual Property Rights
NATO	North Atlantic Treaty Organization
ODA	Official Development Assistance

OECD	Organisation for Economic Co-operation and Development
RED	Renewable Energy Directive
UME	Unidad Militar de Emergencias
UNCLOS	United Nations Convention on the Law of the Sea
UNDP	United Nations Development Programme
UNFCCC	United Nations Framework Convention on Climate Change
UNHCR	United Nations High Commissioner for Refugees
UNSC	United Nations Security Council

1

INTRODUCTION

It is now commonly asserted that climate change will fundamentally change international relations. A huge amount has been written and researched on the incipient physical impact of climate change. Differences persist over the scale of such effects, but the question is exhaustively analysed. The impact such change is likely to have on the broader challenges of international security is harder to foresee. Predictions are plentiful that global warming will increase conflict within and between states; intensify food insecurity; menace the global trading system; and unleash waves of migration. Many scenarios are painted of climate-induced apocalypse. A concern commonly drawn out in such accounts is that climate impacts will trigger non-linear events, through so-called feedback mechanisms. The challenge lies not only in adjusting to a hotter climate but in preparing to react to unpredictable changes that could feed into geopolitics in a frighteningly precipitous fashion. What this means for the conduct of international relations is profoundly uncertain. It is a question that requires systematic analytical attention and assessment.

Governments are beginning to incorporate the warnings into their foreign policy initiatives and instruments. It is this emergent 'climate security' agenda that is investigated in this book. Debates need to move beyond the many warnings of dire consequences ahead; of droughts, uninhabitable swathes of land, climate-induced conflict, mass migrations, food shortages and a general sense of heightened global friction. It needs rather to assess in analytical depth what the impact is on concrete strategies of security and to look in finer-grained detail at the appropriateness of governments' nascent responses.

This book is about the impact of climate change on European Union (EU) security policy. The focus is on actual policy rather than merely outlining possible future scenarios. Many books and articles have been written assessing how far climate change needs to be conceived as a security question. But few have attempted to chronicle in detail how governments are reconfiguring their geo-strategy in response. The excellent volumes covering the EU's role in international climate diplomacy still tend to omit any direct analysis of its role more specifically in relation to climate security.[1] Moreover, while the extensive literature on climate change has incorporated a focus on climate-induced conflict and associated debates over adaptation measures, this book goes beyond this standard focus to assess the broader international relations of climate change. I depart from the view that climate change can no longer be deliberated as a separate area of policy but must be incorporated into the more mainstream debates pertinent to EU common foreign and security policy (CFSP). I ask: what in the broadest sense is the EU's 'climate security' policy?

The EU and its member states have formally acknowledged the broader political ramifications of climate change for over a decade. Of course, the Union is widely seen to have taken up the core, mainstream dimensions of the climate agenda with more serious intent than most other countries and regions. Yet, I find that the tangible, climate-sensitive changes to EU *foreign and security* policies may have many admirable aspects but are not yet extensive. Moreover, changes that have been made exhibit a mix of different strategic philosophies that detract from the clarity and effectiveness of EU initiatives. This book's key contribution to on-going analytical debates is to look at the level of concrete *policy* adaption; in doing so it finds that while many aspects of EU foreign policies have begun to change they do not yet accord 'climate security' unequivocal or sufficient priority.

Vital questions

The definition of climate security remains fluid and contested. It is defined here as the use of all means – diplomatic, financial, economic, military and technical – to address the security-related challenges presented by climate change. It is widely agreed that climate security requires a broad definition that encompasses human, internal and international security.[2] This book draws from the widely advocated 'nexus approach', a definition which argues that climate change essentially challenges the interstices between markets, resource management, military power and human security.[3] In this sense, climate security is categorically not synonymous with militarisation but understood in a broad sense: climate change is likely to affect traditional, defence-oriented definitions of security, but also the broader determinants of stability such as patterns of economic development, alliance-building, the spill-over impact of civil conflicts, migration flows and shifts in relative power balances. This definition seeks a balance: climate security refers to a wider range of policy questions than those of a purely military nature; but it also requires a narrower specification than simply enumerating the physical ramifications of global warming per se.

Official Western definitions seem to give credence to these broadly cast definitions. The US has influentially defined climate change as an 'accelerant of instability'.[4] The EU defines it as a 'threat multiplier' – that is, as one factor that reinforces many of the existing tensions that present challenges for European security, of both a traditional and non-traditional kind.[5] The book will unpack the different dimensions of this definition and measure the EU's policies against its own conceptualisation of climate security.

In doing so, the book speaks to a lively and on-going theoretical debate about the nature of European foreign and security policies. I understand EU strategies in the broad sense of those pursued collectively through European Union initiatives and instruments, alongside those developed nationally by member states; the relationship between these two levels is one factor carefully examined through the different chapters. The book's conceptual framework counterpoises liberal, cooperative approaches to climate security against more realist, self-help notions of strategic interest. It is pertinent to study the EU as an actor in the climate security debate precisely because it has been in the vanguard of designing more holistic approaches to security. This gives its deliberations over nascent climate security challenges a resonance in the broader global community too. The EU's hitherto lead role in international climate policy would seem to place it well for influencing a more tightly defined climate security agenda. Weaknesses in the EU's response would have serious implications both for EU 'actorness' and the nature of the reshaped international system.

The scope of policy questions properly related to climate security remains a topic to be investigated. Coverage of 'the international politics of climate change' is still invariably taken to refer to the reasons why the United Nations (UN) has not delivered a binding agreement on global emissions.[6] The literature on 'climate policy integration' has burgeoned but this concept has not been applied to EU *foreign* policy as such.[7] Projects such as that of the UK's Foresight centre on the international implications of climate change still define 'foreign policy impacts' in a relatively traditional sense of global warming's consequences for economic growth, health costs and the UN's ability to deliver climate solutions. Such studies have advanced the strategic debate but remain predictive and prescriptive; they are still in the mode of warning what consequences might follow and urging governments to 'do something' to prepare. On their own admission, they do not offer an inventory of measures being adopted or of how foreign policy is actually adapting.[8]

Against this background, the book seeks to advance debate by focusing on two over-riding questions:

First, beyond the now well-established rhetoric, how far is climate security actually prioritised within EU foreign policies?

Second, where the EU has upgraded its climate security policies, in which direction has it moved: has it intensified holistic and cooperation-based approaches or reverted to increasingly defensive and traditional notions of security?

My guiding hypothesis is that we should expect to see an increasingly close match between the EU's established foreign policy identity and its incipient approach towards climate security. This is the case because the holistic and interconnected challenges posed by climate change magnify precisely those strategic principles that ostensibly form the foundation of the EU's long-standing and distinctive security identity. The book scrutinises how far EU responses in practice accord to an ideal-type liberal–cooperative framework.

Chapter two offers a very brief overview of what international reports, official documents and academic assessments have said in recent years about the likely effects of climate change. The book does not seek to add to this literature, but merely set the context for an assessment of EU security policies. Chapter three conceptualises the EU's *sui generis* identity as a security actor. European foreign and security policy is made up of a complex balance of governance dynamics. It is vital to understand this balance between realpolitik and liberal–cooperative dimensions because it conditions the way in which climate security is operationalised. Understanding EU security governance is essential to explaining the way that EU climate security combines 'realist–rivalry' and 'cooperative' dynamics. In line with this *sui generis* security governance, chapter four provides an account of the overarching evolution of EU climate security commitments. Chapter five explores the complex bearings that the EU's existing climate change and energy policies have had on developing efforts towards a climate security strategy. Subsequent chapters then assess the nature of EU policies in each component part of the climate security agenda: military–defence configurations; conflict resolution strategies; and geo-economic power.

Findings: serious intent, but inchoate in sum

I uncover an impressive range of new substantive EU commitments in the realm of climate security. To some extent, presenting climate as a security issue has galvanised the policy agenda. Climate change has certainly begun to have an impact on the general definition of European security policy. Energy security has traditionally been conceived and dealt with in terms of relations with key oil and gas producers with little geo-strategic dimension evident in climate change adaptation; policy-makers have moved to rectify this omission.

However, it remains doubtful that a fully geopolitical EU strategy of climate insecurity has taken shape and been implemented in practice. While statements, studies and conferences have been plentiful, their impact on actual European policies remains less than far-reaching. If anything, the rate of policy innovation has slowed, as since 2011 the EU has been preoccupied with other pressing priorities; many climate security strategies were introduced in the years up to 2010–11 only for their follow-up momentum to falter. The definitional issue is one with which European policy-makers still grapple uncertainly: climate security still means different things to different parts of the EU decision-making machine. While the EU is today more prepared to respond to punctual, climate-induced emergencies, neither governments, EU institutions nor militaries have readjusted their global alliances and partnerships to pre-empt the broader impacts on international order. The EU has begun to tackle select elements of climate security, but has yet to put in place a full-spectrum climate foreign policy.

Neither is it clear that climate change is pushing European security policies in one clear conceptual direction. A glaring paradox presents itself. Climate change is the most dramatic case of an issue that requires deeper international coordination; but it also provides an incentive for self-help survival strategies. Some of the substantive responses adopted by European governments have shifted policies towards more international cooperation, while others have tilted more to zero-sum rivalry. Formally, most EU commitments accord strongly to liberal security approaches, predicated on positive sum multilateral cooperation. There is little evidence that European governments are inclined towards a strongly militarised approach; policy developments do not (yet) seem to have borne out sceptics' fear that the whole climate security agenda is being manipulated disingenuously by militaries to re-assert 'hard security' pre-eminence. Climate change has not been 'securitised' in the classic sense of engendering antagonistic, exceptional counter-measures; rather, the EU has been drawn to a more varied form of 'securitisation-lite'. Yet, a concern is evident among diplomats that climate insecurity may challenge the liberal–cooperative approach in the long term. The EU has inched in ad hoc fashion towards a balance between state security and human security logics in its climate geo-strategy. So far, European governments are hedging their bets, mixing collaborative and competitive strategies – although the degree of commitment to climate security has not been strong enough to push EU foreign and security policy decisively in one direction or the other.

An increasing area of concern is how climate change affects EU strategies in the area of conflict prevention and resolution. The resources devoted to tempering conflicts have increased and more comprehensive policies have taken shape around climate-related tensions. The stated priority has been more assiduously to address underlying governance pathologies in fragile and developing states, on the grounds that climate stresses render containment-based strategies to conflict even more clearly insufficient. Yet responses to 'climate conflict' remain underwhelming. There has been little willingness to contemplate military missions; resources dedicated to conflict resolution have remained meagre; and EU development policies incorporate relatively few security-related specificities. A notable shortfall is that deliberations over migration policies have shown little willingness to respond in any constructive way to the prospect of increased climate-induced movements of people.

The shift to low carbon technologies is influencing the balance of geo-economic power between states, and the EU has begun to respond to this nascent re-ordering. International rules governing green technology are taking shape: international power balances will be conditioned by who has this technology and under what conditions, who gains access to it and how. Many policy-makers have designed open-market responses in the belief that the shift to a low carbon economy represents an opportunity for the West to recuperate power vis-à-vis China and other powers. At the same time, a more commercially mercantilistic flavour is evident in relation to European positions on green-related intellectual property rights, international investment regimes, rules of procurement and state subsidies. Tensions between internal renewables development and external cooperation in this area are increasingly apparent. While it would be an exaggeration to suggest that this has engendered conflictual international relations, an incipient sense of rivalry and competition overlay more cooperative EU renewables projects with non-European states.

Many predict that the rising salience of climate security will push the world towards more multivariate forms of governance, as desired outcomes will require networks involving governments, private companies, civil society actors and international organisations. As well as more numerous actors, more experimental and looser processes of governance will be apposite. There is indeed some evidence that the climate agenda has ushered in a slightly more inclusive structure of EU security governance. Policy-makers concur that the politics of climate security must be understood through the lens of new forms of public–private governance, and a plethora of initiatives have been designed to facilitate this trend. However, the influence of such networks in concrete policy decisions remains limited. Civic organisations and businesses have themselves been ambivalent and uncertain over the climate security agenda. It cannot yet be said that their influence is fully factored into concrete EU geopolitical thinking and initiatives. Although a wider net of actors has been included in consultations, it is premature to talk of climate concerns reshaping the qualitative nature of EU security governance.

The book's concluding chapter expands on these findings and passes more normative judgement on incipient EU climate security policies. I argue that the risk appears less one of the EU having securitised climate change too heavily, but rather of insufficient linkage having been made in practice between on-going European policies and impending climate-related strategic challenges. The EU still needs to attach greater priority to deeper international cooperation in pursuit of collective security and qualitative change in the nature of economic growth. It needs to map out more sharply defined policies on these matters as part of its mainstream, day-to-day foreign policies, in addition to current efforts within areas of ring-fenced 'climate crisis management'. All agree that dealing strategically with climate change's consequences will be uniquely challenging, especially as unpredictability will be a key feature, militating against clear geo-strategic doctrine. While the EU has made much progress, it still requires a vastly improved set of instruments and partnerships to help manage the geo-strategic, and not only physical, turmoil that climate change has in store.

Notes

1 For one of the strongest overviews of EU climate leadership in general, see R. Wurzel and J. Connelly (eds), *The European Union as a Leader in International Climate Change Politics*, London, Routledge, 2011.
2 A. Maas, *Shifting Bases, Shifting Perils*, Berlin, Adelphi Research, 2010, p. 13.
3 Transatlantic Academy, *The Global Resource Nexus: The Struggles for Land, Energy, Food, Water and Minerals*, Washington, Transatlantic Academy, 2012.
4 Department of Defense, *Quadriennial Defence Review 2010*, www.defense.gov/qdr/images/QDR_as_of_12Feb10_1000.pdf, p. 85.
5 EU High Representative and the European Commission, *Climate Change and International Security*, Brussels, European Union, March 2008.
6 For example, in the authoritative D. Helm and C. Hepburn (eds), *The Economics and Politics of Climate Change*, Cambridge, Cambridge University Press, 2009.
7 C. Dupont and S. Oberthür, 'Insufficient climate policy integration into EU energy policy: The importance of the long-term perspective', *Journal of Contemporary European Research*, 8/2, 2012, 228–47.
8 A. Sentance and R. Betts, 'International dimensions of climate change', *Climate Policy*, 12/1SUP:01, 2012, 1–5.

2

THE IMPACT OF CLIMATE CHANGE ON SECURITY

This introductory chapter outlines how experts suggest that climate change will magnify a broad range of existing security threats. It summarises the views of a large number of reports, documents and articles which argue that climate change turbo-charges a plethora of current security challenges and gives rise to qualitatively new types of threat that do not fit into established security frameworks.

Pleas for a better assessment of the geopolitics of climate change are not new. For some time, academics have included environmental security within a broadened concept of non-traditional security.[1] As far back as 1987 the Global Commission on Environment and Development called for environmental impacts to be factored into an expanded notion of security.[2] Twenty-two years later the United Nations (UN) secretary general's 2009 report, *Climate Change and its Possible Consequences for Security*, predicted that climate change would 'challenge the existing system of international security governance'; and it lamented that deliberation was still urgently required on how international security policies should react to global warming's deleterious impact on conflict, human rights, migration and basic food supplies.[3]

Meghan O'Sullivan has recently pointed out that the focus of debate has been too narrowly on the technology of the low carbon shift, while we should also be thinking about 'how such a shift will change the world in which we live – in ways far beyond simply how we fuel our cars or make our electricity, but also in the sorts of international security challenges we will face'.[4] Other experts lament that the climate change community still remains dominated by economists, with few conceptual links to the terrain of political scientists or international relations experts.[5] This book aims to make a modest contribution towards rectifying this situation.

Rising impacts

Academic studies, official research consortia and several think-tank projects – including those run at the Woodrow Wilson Center,[6] the Center for American Progress[7] and the Royal United Services Institute[8] – have all compiled extensive data on the probable effects of climate change on security considerations. At the end of 2012, the American Security Project published *Climate Security Report*, which argued that the range of threats had intensified and accelerated appreciably in recent years.[9] The Fifth Assessment Report of the Intergovernmental Panel on Climate Change (IPCC), published in September 2013, offered the first assessment of climate change effects for six years, and reinforced the message that effects were beginning to be profound. The report provided a more fine-grained assessment of differential impacts in different regions, although admitted that models still could not get to country-level predictions.[10]

The most common prediction is that climate change will intensify the frequency and magnitude of natural disasters. The landmark Stern review predicted that 200 million people globally will be affected by climate-driven droughts and floods.[11] The *Climate Vulnerability Monitor* calculates that climate-induced natural impacts already claim 350,000 lives a year and render 50 countries acutely vulnerable, and estimates that a business-as-usual scenario would leave 132 states suffering high vulnerability and cause 1 million deaths a year by 2030.[12] In 2012 the IPCC argued that these fluctuations have become more extreme and difficult to predict.[13] The portion of the IPCC's Fifth Assessment Report published in March 2014 places particular stress on the security risks of climate change.[14]

The *Global Climate Risk Index* has ranked Honduras, Myanmar, Nicaragua and Bangladesh as the world's most vulnerable countries; statistics since 1992 also show Portugal (20th), Italy (24th) and Spain (32nd) high on the list of countries at risk.[15] Global Humanitarian Forum also argues that heat-waves, floods, storms and forest fires will be responsible for as many as half a million deaths a year by 2030, 310 million people suffering adverse health consequences and 20 million more falling into poverty.[16] The Red Cross has called for early warning systems, more extensive flood-walls, flexible and drought-resistant agriculture and provisions for refugees; there is, it argues, a need to 'build climate risk management into all our decision-making'.[17]

One of the main drivers of climate-induced conflict will be food. The International Food Policy Research Institute projects an increase of between 10 and 20 per cent in the number of people at risk of hunger as a consequence of climate change. By 2025 nearly 1.5 billion people will face crop scarcity and 2 billion water scarcity.[18] International Alert director Dan Smith argues that recent price rises have cancelled out a century's advances in food accessibility and in terms of global stability dwarf the impact of the post-2008 financial crisis.[19] Chatham House's 2013 *Resources Futures* report stresses the precarious nature of global interlinkages that have begun to magnify the effect of climatic stresses manifest in poor harvests; for instance, it concludes that climate change is one factor behind a 'new normal' of intense resource price volatility.[20] The United Nations Development Programme (UNDP) calculates that water stress already kills more people each year than violent conflict; it points to growing tensions in the management of major river basins especially in Asia and the Middle East.[21]

A further widely expected consequence will be a huge increase in migratory flows. In 2011, UN high commissioner for refugees, António Guterres, warned that climate-induced migration was 'the defining challenge of our times'.[22] Migration will increase either way: if developing countries do take climate change seriously many millions may lose jobs in traditional, high-polluting sectors; if climate measures are not taken, scarcity and climate stress factors will drive people out. Water shortages in North Africa are already driving migration into Europe. Predictions of climate-induced migrants by 2050 vary considerably: Christian Aid estimates 1 billion, both the International Organisation for Migration and the Stern review more cautiously estimate 200 million, while the United Nations High Commissioner for Refugees (UNHCR) offers a fairly indecisive spread of 250 million to 1 billion.[23] The UNHCR estimates that the nine out of ten natural disasters that are climate-related now displace 6 million people each year. People will move both within countries and across borders.[24]

Many different terms are still employed: environmental migrant, environmental refugee, environmentally displaced person, climate refugee. This reflects disagreement over exactly what will drive such migrants and how they should be treated in a legal sense. While the precise causes of such large flows are still conjectural, they are generally seen as likely to emerge from a nexus of triggers, with migration adding to a general instability already latent in many regions.[25] It is impossible to separate out climate from other types of migration; climate is one amongst several drivers that combine in complex variation. Climate change will increase the demand to move but may increase poverty and other obstacles to people actually being able to relocate; this will increase local strains and tensions.[26]

Climate change is likely to be an accelerator of state fragility. A 2003 Pentagon report predicted war, starvation, disease and institutional collapse within a large number of states, including China.[27] The Central Intelligence Agency (CIA) opened a Center for Climate Change and National Security in 2009 with predictions of dramatic conflict dynamics.[28] In late 2010, Ban Ki-moon urged further action on climate change mitigation, stating a UN finding that 40 per cent of internal conflicts are now due to environmental exploitation and resource scarcity.[29] In 2012, ten UN missions were climate-related. The International Crisis Group predicts that conflict will arise from scarcity but also increased climate variability that may cause short-term economic shocks, reduced employment opportunities and increased recruitment to armed groups.[30] International Alert insists that a total of 46 nations and 2.7 billion people are now at high risk of being overwhelmed by armed conflict and war because of climate change. A further 56 countries face political destabilisation, affecting another 1.2 billion individuals.[31]

James Lee distinguishes between sustained periods of divergent climate patterns acting as a background factor to conflict; climate change interacting with other, intervening variables; and the more proximate triggers required to set off strife. He argues that 'Hot Wars' will erupt in an Equatorial Tension Belt, due to desertification; a 'transition zone' moving from being temperate or tropical to becoming semi-arid; and other countries moving from being 'tropical forest to tropical grassland' and thus having to readjust economic and livelihood systems. All such dynamics can be termed 'livelihood wars'. This will tend towards a militarisation of resources. Internationally, the threat of climate change will become a justification for preventative war: conflicts perpetrated to stop another country from taking harmful environmental actions that impact another, to prevent countries from depriving others of their fair share of environmental resources and in order to acquire basic resources.[32]

One comprehensive report on water scarcity in the Levant describes how this area already stands as one specific microcosm of what the future is likely to hold more generally. Tensions over water management have long plagued negotiations in the Middle East and been the subject of periodic social flare-ups. Most key water sources in the region are trans-boundary, with the largest water aquifers shared between two or more states and territories. Severe droughts from 2008 have been linked to the region's worsening political climate. Climate change here is both cause and effect of rising tensions. Most obviously it increases the premium on retaining control over dwindling water sources. It makes governments less likely to cede land in the name of peace accords and increases the risk of pre-emptive resource seizures. It is one factor underpinning the persistence of patronage-based governance. But the general level of existing political distrust is also a causal effect of failures to cooperate on resource management. Over-inflated military budgets drain resources away from climate mitigation and adaptation.[33]

Climate change is also likely to have an impact on the broader patterns of international relations. Ian Morris argues that no other factor will come close as a determinant of geopolitics, not even China's rise or the devastating post-2008 financial crisis. Political history since the earliest times can, he argues, be explained in part by the relationship between geography and the succession of different models of production. With the capacity ceiling of the current mode of production now imminent, a fundamental change in global politics is likely.[34]

Some writers predict a whole series of tense, new cleavages arising from climate-related redistribution: coastal versus inland cities; north versus south; rich versus poor; and developed versus under-developed countries – these overlain by deeper class conflict as the rich use their means to protect themselves.[35] The Earth Security Initiative fears that richer nations will buy up land and resort to even more agricultural protectionism, and that this will magnify the risk of local conflict and undermine local rights.[36] International inequalities will deepen. A contradiction will take root: the effects of climate change will push governments to want to protect themselves; as the best protection is wealth, they will pursue higher gross domestic product (GDP) growth rates and thus consume more. States will desire stronger buildings, more air conditioning and more expensive storm protection.[37] They will intensify the very features of the economic system from which they require protection.

The Center for American Progress has concluded that climate factors were one of the 'stressors' that contributed to the post-2011 wave of revolts.[38] The late Tony Judt argued that climate change would be part and parcel of a general demise of liberalism:

> Men and women will be thrown back upon the resources of the state. They will look to their political leaders and representatives to protect them: open societies will once again be urged to close in upon themselves, sacrificing freedom for 'security'.

Climate change will play a central role in the revival of the nation-state and national politics, against the dominance of market interdependence, he predicted.[39]

States' adaptability to climate change will condition relative degrees of power. Constance Lever-Tracy points out that climate change is likely to 'throw into disarray prior hierarchies and certainties, promoting new interests and alliances'. States that do not adapt well will fade. Those that successfully develop the various techniques of geo-engineering – such as sulphuric acid spray, artificial clouds, sunshades and carbon sequestering through artificial trees – stand to gain power. Conversely, oil states will become weaker as the move to low carbon progresses.[40] At the end of 2012 a World Bank report warned that climate change's geo-economic impact was being under-estimated, as it was set to extend beyond generating localised conflict and undo the crucial linkages upon which global economic development depend.[41]

In some senses, a swathe of northern states, from Canada through Scandinavia into Russia, could be big gainers from global warming. Not only will they be amongst the few places to benefit from increased rainfall and water supplies, longer growing seasons and negligible population stresses, they also stand to gain from access to new hydrocarbons and trade routes as the Arctic ice retreats. This does not mean Arctic crop yield increases will compensate for losses in the south, or that there will be a mass exodus to the far north that will still be an extremely harsh place in which to live and survive. But the states on the Arctic rim will feel more comfortable while others are under strain.[42] Arctic geopolitics are now on the front line of international politics.

In short, uncertainty over the form that the shift to low carbon will take is likely to ensure an extended period of fluid jostling for position between states and between different national companies. It is argued that observers missed just how far the political impacts of climate change already engendered a fluidity of new alliances at the Copenhagen summit in 2009. Groupings extended beyond simply the rich West versus poor developing states; smaller groupings formed of Arctic states, low-lying islands, unconventional oil states and others.[43]

A final key change is that security will become about more speculative forms of risk mitigation, less about reacting to traditional, substantiated material threats.[44] The over-riding effect may be that of uncertainty. Climate models devised so far are relatively imprecise; we do not really know with certainty what impacts of a warmer climate will be felt where. Feedback mechanisms are by definition difficult to predict. Preparedness and agility in security policy will be at more of a premium. Nature is metastable: it retains stability in the face of multiple and grave disruptions, until the moment when it does lose equilibrium in a really big way. One National Intelligence Council-sponsored study of potential physical and political effects, disaggregated to the country and regional level, concludes that 'easy generalisations about the security implications of climate change will be hard to come by', and that most stress may flow from the expectation of impact rather than the actual physical effects per se.[45] Many fearfully predict little but the unpredictable.

The doubts

While strong consensus now exists on all these likely political consequences, it is not universally accepted that climate change will have a significant impact on global security. Some still doubt the conceptual validity of ranking climate change alongside the more traditional threats to security.

Doubts come in various shades. Some of the most nuanced analysis has suggested that climate change is indeed a security risk but one whose importance should not be exaggerated. In his widely read dissection of the mounting risks to civilisation, Martin Rees cautioned that climate change was a factor of concern but not one of such magnitude that it should displace policy efforts in other areas such as nuclear non-proliferation, the vulnerabilities of technological dependency or uncontrolled scientific advancement.[46] A general argument is that the links between climate factors and security have not been demonstrated carefully enough; debates often list the physical effects of climate change and then an assumption is 'tagged on' that security implications will inevitably follow.[47]

Geo-strategic foresight exercises still struggle to incorporate climate change in a systematic way: it is difficult to predict geopolitical scenarios resulting from climate change beyond offering general observations on which states will be physically hardest hit by global warming.[48] One common line of argumentation is that climate change represents a serious challenge for the adaptability of individual and community behaviour but should not be conceived as a macro-scale threat to national security.[49]

The mainstay critique has been that Western powers have begun to define climate change as a standard security threat in military terms, when it is far from being comparable to such challenges and consequentially risks being pursued in a way that aggravates conflict. Climate security is being defined as a series of threats coming from the 'periphery' to destabilise the Western 'centre' in traditional hard security fashion, when the danger is less about security than Western overconsumption menacing the 'periphery'. As a result, climate impacts are 'not obviously a matter of 'national security' as normally defined.[50]

Climate factors may add one additional source of stress on already fragile states, but it is not yet certain that they will trigger qualitatively new conflict dynamics in currently stable states; many of those states likely to face the more serious effects actually have institutions with reasonably strong 'adaptive capacity'.[51] According to this line of thought, climate will not directly cause conflict but merely intensify possible 'pathways for instability', to the extent that it awakens pre-existing tensions.[52] Collections of peer-reviewed studies exhibit very different expert views on any causal link between climate change and conflict – at least some of these experts suggest that framing climate change as a driver of conflict is becoming a dangerously self-fulfilling prophecy, as governments militarise in preparation for an eventuality that for the moment remains uncorroborated.[53] A World Bank study has also cast doubt on any empirical link between climate change and conflicts.[54] Sceptics are left asking whether militaries have stressed the need for a security dimension to climate change mainly as a way of justifying their own hold on scarce finances.[55]

More radical climate sceptics insist that conflict is more likely to result from the harsh economic sacrifices imposed in the pursuit of ill-founded and arbitrary emissions targets than from any climate change itself. Steps being taken in the name of suppressing a misguided notion of climate conflict in poor states actually *clash* with broader security interests. Claims that climate change represents a more serious threat than terrorism have, doubters say, percolated onto the agenda due to the influence of powerfully placed environmentalists and not as a result of any measured threat assessment.[56]

Another example of doubt is found in the area of migration trends. The German Marshall Fund finds that large-scale forced or distress migration related to environmental change is likely to be 'internal, regional, and temporary'.[57] The complexity of migration makes it extremely difficult, if not impossible, to single out the environmental factors of displacement. This makes any estimate of the number of current and potential environmentally induced migrants highly contestable. The Norwegian Refugee Council warns of disingenuous political agendas exaggerating the spectre of environmental migration for their own purposes.[58] Some writers suspect companies of playing up the climate security agenda as a means of accessing 'green subsidies', and political elites of wanting to fashion an institutional framework of global governance that gives governments more power and negates local democracy.[59]

Some experts insist that policy should focus on traditional energy security concerns as attempts to broaden debate out to a climate-inclusive concept of security simply produce a tangle of tenuous assertions that muddy policy calculations. They question the easy assumption that energy security and climate change policies are mutually reinforcing simply because renewables usage makes states more independent. Rather, prioritising climate change goals can undermine other aspects of energy security – increasing dependence on gas, renewable sources and nuclear energy. Indeed, these experts fear that climate advocates now use the 'security threat' discourse as a means to push governments to adopt more green policies. So, they infer, the disingenuousness runs both ways: the climate community is using the security community, and vice versa.[60]

Well-respected experts insist that the geopolitics of energy will continue to be about the politics of oil and gas, not renewables or climate impact.[61] Some explicitly advocate a prioritisation of energy security even at the cost of slower progress on climate security – with energy and climate security seen as mutually exclusive trade-offs, not an integrated whole.[62] Many experts argue that the real game changer is the discovery of shale gas. This has already largely eliminated the US's dependence on external hydrocarbon supplies. Since 2011 gas output has doubled, placing downward pressure on prices. The geopolitics of energy is, many argue, increasingly a question of the effects of a shift to shale gas, far more than of the prospective rise of renewable sources. Access to shale gas will become the key dynamic from a security perspective, not a search for independence based on renewables.

Conclusion

For at least two decades now, analysts, non-governmental organisations (NGOs) and international institutions have been warning that climate change is likely to have far-reaching geopolitical effects. A wealth of predictions and data has accumulated pointing to the expected severity of extreme weather events, inter- and intra-state conflict and migration waves. Crucially, analysts have increasingly drawn attention to the more indirect changes that are also possible to the general contours of international relations. Climate change has increasingly been recognised as a likely proximate cause of conflict and tension, but also an underlying shaper of the way in which international relations are conducted. A striking breadth of myriad political, social, economic and strategic impacts is now traced back to climate change. Taking on board the full implications of all this analysis invites consideration of how climate security is being dealt with at all levels of European Union (EU) foreign and security policy.

All this informs the challenge for EU policy. The wide-ranging consensus is that the effects of climate change are likely to be deep and broad enough to require priority attention not just in terms of direct climate negotiations but across the gamut of foreign policies. From the often bewildering litany of predictions, it is useful to extract a concise number of possible outcomes at the level of concrete policy responses. This provides us with a more operational template, against which to evaluate EU policies in the chapters that follow:

Prediction: More violent conflict, driven by climate-related stresses; increased frequency of natural disasters.
Policy challenge: To prepare for climate-related military missions and conflict prevention.

Prediction: More acute tensions across borders.
Policy challenge: More structured diplomacy and incentives focused on climate-related global dynamics.

Prediction: Resource scarcity and economic stresses.
Policy challenge: Economic and development policies broadened to temper such risk.

Prediction: Climate change as additional cause of shifts in global power.
Policy challenge: To deal with and pre-empt new patterns of international alliances.

Predictions: Stresses to democracy.
Policy challenge: To move away from state-centred approaches to security.

Prediction: Unpredictability.
Policy challenge: More agile response mechanisms at all levels.

There are, of course, discordant voices that question the geo-strategic reading. In a sense, the doubters make it even more necessary to assess actual policy reactions. It might be said that the EU faces the challenge of avoiding two opposite risks: the first, of under-reacting; the second, of over-securitising climate change. It is this kind of difficult balance that this book also sets out to investigate, as the EU grapples with the climate security agenda. Jeffrey Mazo points out that the academic literature has stagnated at the stage of issuing dire warnings and generic predictions, and that more detailed assessments of concrete policy challenges in particular locales are still awaited.[63] This chapter demonstrates both that there is a long-standing claim for the geopolitics of climate change to be treated more seriously, but also that guidance remains uncertain over what this means in concrete policy terms.

Notes

1 B. Buzan and L. Hansen, *The Evolution of International Security Studies*, Cambridge, Cambridge University Press, 2009, p. 259.
2 United Nations, *Report of the World Commission on the Environment and Development: Our Common Future*, Document A/42/427, United Nations, 1987.
3 United Nations, *Climate Change and its Possible Consequences for Security*, New York, United Nations, 2009.
4 M. O'Sullivan, 'The entanglement of energy, grand strategy, and international security', in A. Goldthau (ed), *The Handbook of Global Energy Policy*, Oxford, Wiley-Blackwell, 2013, p. 43.
5 A. Michealowa, 'An outsider view of climate politics', *Climate Policy*, 11, 2011, 832–33, p. 832.
6 www.wilsoncenter.org/program/environmental-change-and-security-program.
7 www.americanprogress.org/projects/climate-migration-security/view/.
8 www.rusi.org/globalsecurity/climatesecurity/.
9 C. Foley and A. Holland, *Climate Security Report*, Washington DC, American Security Project, 2012.
10 Intergovernmental Panel on Climate Change (IPCC), *Climate Change 2013: The Physical Science Basis*, Fifth Assessment Report, Geneva, IPCC.
11 Her Majesty's Government, Lord Stern, *The Stern Review, The Economics of Climate Change*, London, The Stationery Office, 2006.
12 DARA–Climate Vulnerable Forum, *Climate Vulnerability Monitor 2010*, Madrid, DARA, p. 15.
13 IPCC, *Managing the Risks of Extreme Events and Disasters to Advance Climate Change Adaptation*, Geneva, IPCC, 2012.
14 Intergovernmental Panel on Climate Change, *Climate Change 2014: Impacts, Adaptation, and Vulnerability*, Fifth Assessment Report, Geneva, IPCC, 2014.
15 S. Hammeling and D. Eckstein, *Global Climate Risks Index 2010*, Bonn, Germanwatch, 2013, pp. 21–22.
16 Global Humanitarian Forum, *The Anatomy of a Silent Crisis*, Geneva, Global Humanitarian Forum, 2009.
17 ABC News, 17 June 2009.
18 International Food Policy Research Insitute, *World Water and Food to 2025*, Washington, International Food Policy Research Institute, 2002.
19 D. Smith, 'Resources – the coming crunch and some things that could be done about it', *International Alert Blog*, 7 January 2013.
20 B. Lee, F. Preston, J. Kooroshy, R. Bailey and G. Lahn, *Resources Futures*, London, Chatham House, 2013, p. 4.
21 United Nations Development Programme (UNDP), *Human Development Report 2006, Beyond Scarcity: Power, Poverty and the Global Water Crisis*, UNDP, 2006.
22 Speech at the Nansen Conference on Climate Change and Displacement in the 21st Century convened by the Norwegian Government, 6–7 June 2011, Oslo.

23 IOM, *Migration and Climate Change*, IOM Migration Research Seminars 31, Geneva, 2008.
24 UNHCR, 'Climate change is a humanitarian problem', leaflet. www.unhcr.org/4937fc712.html.
25 M. Werz and L. Conley, *Climate Change, Migration and Conflict: Addressing Complex Crisis Scenarios in the 21st Century*, Washington, Center for American Progress, 2012.
26 Foresight, *Migration and Global Environmental Change, Final Report*, London, The Government Office for Science, 2011, p. 22.
27 Department of Defense, *An Abrupt Climate Change Scenario and its Implications for US National Security*, Washington, US Department of Defense, 2009.
28 http://climateandsecurity.org/tag/cia-center-for-climate-change-and-national-security/.
29 www.un.org/apps/newsFr/storyF.asp?NewsID=23598&Cr=conflit&Cr1=.
30 www.crisisgroup.org/en/key-issues/climate-change-and-conflict.aspx.
31 International Alert, *A Climate of Conflict*. London, International Alert, 2007.
32 J. Lee, *Climate Change and Armed Conflict: Hot and Cold Wars*, London, Routledge, 2009.
33 O. Brown and A. Crawford, *Rising Temperatures, Rising Tensions: Climate Change and the Risk of Violent Conflict in the Middle East*, Winnipeg, International Institute for Sustainable Development, 2009.
34 I. Morris, *Why the West Rules – For Now*, London, Profile Books, 2010.
35 G. Dyer, *Climate Wars*, Oxford, One World, 2010, p. 229.
36 A. Levitsky, *The Land Security Agenda*, London, Earth Security Initiative, 2012.
37 M. Klare, *Rising Powers, Shrinking Planet: The New Geopolitics of Energy*, New York, Holt, 2008, p. 8.
38 C. Werrell, F. Femia and A.-M. Slaughter (eds), *The Arab Spring and Climate Change*, Washington, Center for American Progress, 2013.
39 T. Judt, *Ill Fares the Land*, London, Penguin, 2010, p. 9 and p. 195.
40 C. Lever-Tracy, *Confronting Climate Change*, London, Routledge, 2011.
41 World Bank, *Turn Down the Heat: Why a 4 Degree World Must be Avoided*, Report for the World Bank by the Potsdam Institute, Washington, 2012, p. xviii.
42 L. Smith, *The New North: The World in 2050*, London, Profile Books, 2011.
43 H. Winkler and J. Beaumont, 'Fair and effective multilateralism in the post-Copenhagen climate negotiations', *Climate Policy*, 10, 2010, 638–54, p. 640.
44 E3G, *Degrees of Risk: Defining a Risk Management Framework for Climate Security*, London, E3G, 2011.
45 D. Moran (ed), *Climate Change and National Security: A Country Level Analysis*, Washington DC, Georgetown University Press, 2011, p. 5.
46 M. Rees, *Our Final Century: Will Civilization Survive the Twenty-first Century?* London, Arrow, 2003.
47 C. Abbott, 'An uncertain future: law enforcement, national security and climate change', *FRIDE Comment*, February 2008, 1.
48 One example of this is G. Friedman, *The Next 100 Years: A Forecast of the 21st Century*, New York, Random House, 2009.
49 C. Webersik, *Climate Change and Security: A Gathering Storm of Global Challenges*, Santa Barbara, Praeger, 2010.
50 S. Dalby, 'Climate change and environmental security', in P. Williams (ed), *Security Studies*, 2nd edition, London, Routledge, 2012, p. 314.
51 J. Mazo, *Climate Conflict*, London, Routledge and IISS, 2010.
52 Maas, *Shifting Bases, Shifting Perils*, Berlin, Adelphi Research, 2010, p. 19.

53 N. Gleditsch, 'Introduction: Whither the weather? Climate change and conflict', *Journal of Peace Research*, Special issue on climate change and conflict, 49/1, 2012, 3–9.
54 H. Buhaug, N. Gleditsch and O. Theisen, *Implications of Climate Change for Armed Conflict*, The World Bank Group, Social Development Department, 2008.
55 For a recent and succinct overview of the debate over conflict and climate, see A. Solow, 'Global warming: A call for peace on climate and conflict', *Nature*, 497, 2013, 179–80.
56 C. Booker, *The Real Global Warming Disaster*, London, Continuum, 2009.
57 German Marshall Fund, *Climate Change, Migration and Conflict: Receiving Communities under Pressure?* Washington DC, GMF, 2010.
58 Norwegian Refugee Council, *Future Floods of Refugees: A Comment on Climate Change, Conflict, and Forced Migration*, Oslo, Norwegian Refugee Council, 2008.
59 P. Clover and M. Economides, *Energy and Climate Wars*, London, Continuum, 2010, pp. 160–61.
60 G. Luft, A. Korin and E. Gupta, 'Energy security and climate change: a tenuous link', in B. Sovacool, *The Routledge Handbook of Energy Security*, London, Routledge, 2011.
61 Klare, *Rising Powers*.
62 A. Froggatt and M. Levi, 'Climate and energy security policies and measures: Synergies and conflicts', *International Affairs*, 85/6, 2009, 1141.
63 J. Mazo, 'Cold comfort', *Survival*, 52/6, 2011, 151–60, p. 159.

3

THE EU'S SECURITY IDENTITY

This chapter outlines on-going debates about the European Union's (EU's) general security identity. Clearly, understanding the evolution of this identity is essential to unpacking the EU's responses to the climate security agenda. The chapter describes the EU's *sui generis* foreign and security policy structures, and reports on how its external identity has been affected in recent years by the economic crisis. The EU's complex institutional structures still seem to militate against effective and coherent decision-making, despite several rounds of treaty change. The EU and its member states remain committed in philosophical terms to a basic internationalism in their security thinking; recent policy trends, however, suggest a possible dilution of such proactive and cooperative efforts. Sceptics might ask whether Europe's climate strategy will matter much at all when its relative power in the world is in severe decline. The chapter suggests a framework for how to match the EU's security identity to the strategic challenges posed by climate change. The EU rhetorically stands ready to lead a cooperative-based approach to climate security, but its resources and political will appear increasingly stretched in the realms of both soft and hard power. This impacts on how well positioned the EU is to meet the challenge of climate security, in a context of debates over whether this agenda requires a tilt towards more realist or liberal–cooperative approaches.

A unique security identity?

Academics most commonly argue that the EU has developed a *sui generis* identity in foreign and security policy. This is a product of the Union's unique and complex institutional structures and the historical experience of its own integration process. As is well known, the EU is somewhere between a nation-state and a standard intergovernmental organisation. Its common foreign and security policy (CFSP) is largely, although not purely, intergovernmental. It consists of cooperation between national governments, which pursue their own national diplomacy in parallel with EU cooperation. Coordination commitments in the field of defence are more limited. International trade policy is largely supranational and managed by the Commission; as we will see, the Commission holds significant powers in the fields of development aid and energy and climate policy too.

The EU has long been seen as embodying the epitome of cooperative approaches to security. Most commentators and diplomats define the EU as a multilateral strategic actor, drawn by its very nature to positive-sum and holistic security policies. The EU has regularly repudiated zero-sum power politics. The EU has been defined as a 'collective power' that does not act in its own narrow, selfish interest but considers the general interest of the international system.[1]

As early as 2003 the European Security Strategy laid out an impressively all-inclusive and forward-looking approach to security. This document committed the EU to a strong engagement in the full range of security challenges, through a balanced use of economic, social, political and military instruments. The EU's much-commented-on self-definition as a post-modern security actor lays claim to a conviction that European interests are served not by seeking to maximise relative power but by assisting other states to become stronger, more prosperous and more effectively governed.[2]

The EU stresses that rules-based behaviour is the constitutive cornerstone of its geo-strategic identity. It proclaims soft power to be the most effective form of strategic power. In 2003 Jurgen Habermas and Jacques Derrida famously wrote that European opposition to the invasion of Iraq revealed a firmly rooted and distinctive EU foreign policy identity – one wedded to the principles of cooperation, non-intervention and soft power, and utterly different from the US's international profile.[3]

Together, the Commission and member states are the world's lead development donors. The EU has contributed greatly to expanding the definition of security beyond traditional notions. From an early stage and in all its different regional partnerships, the EU has insisted that security cannot be reduced to traditional, defensive concepts but must be broadened to include soft security questions such as economic development, the combatting of cross-border crime, counter-radicalisation, human security and environmental questions.[4]

It is commonly argued that realist theories cannot explain the progression of European foreign policy cooperation and that a range of other analytical frameworks are better suited to CFSP. Indeed, the latter has become something of a laboratory for emerging international relations theories. Different pathways are emphasised towards unified, cooperative norm-based EU foreign and security policies. Neo-functionalists stress how external positions converge on the basis of internal sectoral commitments. Constructivists talk of common EU policy identities outweighing realist national calculations of narrow self-interest, and the emergence of a distinctively dense area of 'security governance'. Institutionalists argue that decades of regularised processes of coordination have bred common outlooks between member states around rule-based international action.[5]

Some academics insist that these features have become stronger since the beginning of the eurozone crisis in 2009. For all the tribulations of the economic crisis, member states have, they insist, found successful means jointly to manage their economic imbalances. The EU has emerged vindicated as a model for effective multilateralism, in the vanguard of charting cooperative solutions to the problems associated with financial globalisation.[6] Some experts insist that the crisis has prompted the EU to engage in the kind of flexible, multi-level problem-solving that makes it even better suited to navigating the very different diverse strategic challenges of a reshaped world order.[7]

The EU's response to the eurozone crisis has not been purely defensive. There are aspects where the EU has registered improvements in its security strategies during the crisis. The EU has fashioned broader strategic partnerships with rising powers. It has begun to engage more politically in Asia. It has strengthened a number of commitments to human rights, democracy and good governance. It has begun to look outwards at least on some issues, as governments realise that the crisis has left them more in need of international partnerships and interdependent linkages. Some diplomats see a firmer defence of the liberal world order as more necessary now as a means of shoring up European influence against China and other ascendant powers.[8]

Many academics, however, detect changes to the EU's presumed security identity. There is much debate over how far European security policies actually correspond to the EU's self-image. In the context of the economic crisis and shifting global power balances, European foreign policy has struggled to adapt to new international challenges. Some academics argue that in recent years the EU has becoming increasingly wary of entanglement in foreign affairs and reluctant to spend significant resources on security questions.[9] The EU is now more selective in following through on what it always lists as the axiomatic pillars of its foreign policy identity. European support for multilateralism remains notable but has in many areas become more qualified. The means and political will invested in so-called soft security initiatives are in most places and thematic areas woefully inadequate for the magnitude of the tasks at hand. Bilateral trade deals are increasingly prioritised over any attempt to deepen rules-based multilateral economic interdependence. The strict control of migration and surveillance-based internal security policies sit uneasily with the EU's rhetoric of positive-sum 'collective security' with its neighbours and partners.[10]

European support for multilateralism has become slightly more conditional and bounded. One team of academics argues that the EU has gradually adopted a more ad hoc form of multilateralism, varying across issues and with select and varied clusters of partners.[11] If the EU has traditionally held to a strongly institutionalist understanding of multilateralism – the process of international cooperation is valuable in itself and socialises actors around common outlooks – it now judges the merit of international coordination in a more results-oriented, problem-solving manner.[12]

One academic writes of the EU suffering 'normative uncertainty' and shifting towards a foreign policy based on 'bi-multilateralism': its commitment to 'negotiated order' still exists but is now moulded around the pursuit of more immediate and material, national self-interest.[13] The importance attached to geo-economic security appears to have gained relative to other forms of security. While the desirability of positive-sum cooperation and multilateralism is still stressed, European governments have focused more increasingly on relative economic gain.[14]

The 2009 Copenhagen climate summit was a crucial turning point. This seemed to show that a very principled adherence to multilateral principles can easily backfire and leave the EU isolated, if not accompanied by tougher and more contingent negotiating strategies. Noted experts argue that the EU has increasingly used bilateral strategic partnerships with key powers to lever tangible trade-offs en route to deeper multilateral cooperation.[15] Other thinkers concur that member states seek a more direct form of engagement with rising powers that is not predicated so heavily on abstract, EU-like multilateral principles.[16] Security is increasingly seen to require practical and select cooperation through smaller groupings rather than ideal-type, often-unattainable multilateralism. Academics talk of an increasing European preference for minilateralism as a response to more uncertain strategic challenges.[17] The implication of these views is that the EU is becoming more like a traditional foreign policy actor and not primarily one that exerts agency by virtue of itself representing a microcosm of liberal–cooperative security governance.

How united a security actor?

A closely related question is how united EU member state governments are in their foreign and security policy. Overall, EU foreign and security policy embodies the dynamics of rule-based cooperation between member states *and* more traditional, national security policies. This inevitably renders European external policies multidimensional; it is an institutional foundation that, as we will see, deeply affects the nature of climate security policies. The balance between cooperation between member states and divergent national diplomacy has shifted over time and varies between issues.

Many academics insist that the habits of cooperation and shared problem-solving in foreign policy have progressed too far to be easily reversible. Such habits have, they insist, withstood the eurozone crisis well. The EU has developed gradually over fifty years into a strongly embedded 'security community' that the economic crisis in unlikely to undo.[18] Noted theorists argue that even if the crisis weakens certain EU institutional structures it is unlikely to unravel the deeper roots of this European 'security community': peace is ensured though deep interdependence, shared values and the cement of other institutions such as the North Atlantic Treaty Organization (NATO).[19]

Indeed, prominent academics detect a broad resuscitation of neo-functionalist dynamics, with the crisis pushing forward the 'integration frontier' across a whole range of issues.[20] The history of European integration shows that unity tends to solidify in moments of crisis and that the EU has been capable of re-inventing the rationale for cooperation.[21]

Moreover, the reshaped world order requires the kind of pragmatic joint management of new challenges that will bring EU governments closer together. Thus, some theorists insist that there is both a remoulded structural realist motive for deeper coordination, and an internal self-sustaining momentum to deeper foreign policy convergence exogenous to and counteracting the eurozone crisis.[22]

In fact, formal proposals for deepening the institutional structures of foreign and security policy cooperation have become more prominent in recent years. Proposals for political union have included plans for a fully supranational foreign and security policy. A report produced by eleven foreign ministers in September 2012 marked a notable attempt to map out a vision for a more integrated CFSP, along with a European army. At the end of 2012, France, Germany, Italy, Spain and Poland agreed to support a new 'military structure' to oversee common European conflict resolution missions.[23]

Foreign and security policy is one of relatively few areas where the UK remains relatively strongly committed to productive engagement at the EU level. Indeed, for London cooperation on foreign policy may be seen as necessary to help compensate for the UK's marginalisation from the core economic union. In turn, Germany and France have looked for areas to keep the UK from drifting too far apart; foreign and security policy is the most obvious possibility.

Many other observers, however, observe growing disunity. Former German foreign minister Joschka Fischer laments that EU foreign policy today suffers from creeping 'renationalisation' and 'provincialism'.[24] Charles Kupchan writes of the 'renationalisation of European politics' that has distanced the EU even further from 'the collective governance that Europe needs to thrive in a globalised world'.[25]

Intergovernmentalists argue that cooperation between member states is less deeply rooted and occurs on the basis of governments' rationalised calculations of self-interest. Realist writers interpret the euro crisis as the manifestation of broader underlying global political dynamics: in the multipolar word there is no single, overwhelmingly threatening power against which the EU needs to 'balance', removing the main driving force for integration and gradually weakening the glue binding member states together in the most general strategic sense.[26]

Paradoxically, the Lisbon Treaty seems to have produced a shift in the centre of gravity of foreign policy-making back to national capitals and hence more defensive reneging on unity commitments. Critics associate this with a move away from proactive internationalism. The External Action Service (EAS), the new diplomatic body set up under the rubric of the Lisbon Treaty, has focused mainly on classical diplomacy more than inter-issue linkages. Member states now tend to formulate national positions without coordinating with their partners. Germany, France and the UK perceive themselves to have greater room of national manoeuvre in a world of middle powers and shifting alliances around the world.[27]

Security and defence: limitations

In debates about the EU's international identity and presence, the place of security and defence policy has long been viewed critically. A repeatedly expressed concern is that European military capabilities have not been harnessed to coherent security strategies and visions. The EU ritually insists that its strategic comparative advantage lies in its ability to manage an increasingly widened concept of security. In reality in most cases the EU has struggled to dovetail the panoply of its soft and hard power instruments within a comprehensive approach to geo-strategy. Experts argue that European security policy has not been good at linking its generic set of strategic principles to an appreciation of where concrete features of geographical change, in specific places, impact upon supply imperatives.[28] The EU is good at focusing on the generic way it prefers to do international relations, but has failed to develop 'grand strategy' in the more concrete sense of where and how its power resources are to be deployed with security interests in mind.[29]

In the midst of the deficit-reducing 'age of austerity' European governments have found it difficult to make the case to their populations and parliaments for hefty defence budgets aimed at long-term threats. The precarious state of the euro and the whole internal agenda of European integration have made external defence and security commitments seem less pressing. European defence budgets have been dramatically reduced. The overall EU defence spend decreased by just over 10 per cent from 2008 to 2012, and reductions of a further 25 per cent are in the pipeline up to 2020.[30]

Financial constraints have encouraged European governments to cooperate in defence capacity-building. The so-called 'burden sharing' agenda has advanced in recent years. In February 2012 the UK and France signed an agreement to deepen defence cooperation, through an integrated maritime fleet, joint exercises in the Mediterranean and inception of a joint expeditionary force by 2016. Germany and Italy have also signed a memorandum on procurement cooperation. Smaller groupings of Nordic states and the Benelux and Visegrad groups have similarly deepened such cooperation. The 2010 Ghent framework has given a coordinating umbrella for these various pooling and sharing initiatives. NATO's Smart Defence initiative has taken the trend further. In 2012 the European Defence Agency was given a strengthened statute, and the defence sector was opened to more single market discipline in 2009.[31] At the end of 2012 member states agreed a new Code of Conduct for Pooling and Sharing. A European Commission Task Force on Defence has explored ways of removing barriers to cooperation between defence firms.

The December 2012 EU summit charged the European Commission and the European External Action Service with drawing up proposals to strengthen the EU's security policy and defence capabilities. Leaders committed to discussing defence cooperation at a summit in December 2013, for the first time since 2008. This summit focused on cost-saving capability-sharing, with leaders agreeing to develop jointly drones and satellite surveillance and to draw up a maritime security strategy; its conclusions make cursory reference to common security and defence policy (CSDP) being relevant to energy security and to a putative EU 'security of supply' regime.[32]

However, the EU's preparedness to intervene effectively to resolve conflicts has been increasingly questioned. Some observers worry that the EU has moved progressively further from its commitment to liberal–cosmopolitan security policies. Development aid has been reduced to conflict-ridden states. One eminent security expert notes that the whole notion that European states can 'do something' to mitigate conflict in unstable parts of the world has 'lost momentum'.[33] It is doubtful, argue some academics, that European policy-makers or publics have come to accept the legitimacy of anything other than very limited and narrowly defined humanitarian intervention.[34]

No pioneers have stepped forward to advance the flexible defence cooperation foreseen in the Lisbon Treaty. The External Action Service itself laments that crisis response mechanisms still need to improve to give substance to the EU's proclaimed 'comprehensive approach' to conflict resolution.[35]

Three new CSDP missions were deployed in 2012: to enhance maritime capacity around the Horn of Africa; to protect an airport in South Sudan; and to train security services in Niger. Yet these were all extremely modest missions, comprising mainly security sector trainers. In Libya, CSDP was absent. The EU priority is to outsource post-conflict peace-building to regional organisations, such as the African Union. Only France stood ready to intervene militarily in any significant way after the Assad regime's alleged use of chemical weapons in August 2013. The French government has pushed for the creation of an EU fund to finance military interventions in crisis situations; other member states for the moment remain unenthusiastic.

Indeed, France has become more active, deploying missions to Côte d'Ivoire, Libya, Mali and the Central African Republic. In each case, however, the military mandate was drawn relatively tightly and France eschewed high-level involvement in long-term peace-building. Germany was sharply rebuked by Paris for refusing even modest pleas for support in Mali. French troops admirably helped stabilise Mali's internal security but moved to hand over quickly to an African-led mission. Overall, the number of EU troops deployed in the field nearly halved between 2008 and 2012.[36]

Sobering experiences in Iraq and Afghanistan have left the UK army cautious about further deployments and uncertain about its whole conflict resolution doctrine.[37] The British army has moved from a focus on enduring campaigns to preparing for a range of less direct contingencies. It will concentrate instead on training foreign armies better to manage their national security and supporting civil authorities to protect infrastructure at home. Experts observe a switch away from offensive armed forces to homeland security, eviscerating stabilisation capabilities beyond its borders.[38]

Differences between member states have made it impossible to update the 2003 European Security Strategy. A process to draw up a new European Global Strategy was launched in 2012 by Poland, Sweden, Italy and Spain, but other member states have not supported the publication of a new, formal strategy.

What link to climate security challenges?

In short, the climate security challenge presents itself against the backdrop of a fluid juncture for European security identities. The question that presents itself is what kind of security identity is most appropriate for tackling the strategic impacts of climate change. Does climate security require the kind of liberal internationalist approach long associated with EU policies? Or does it enjoin governments to adopt more realist self-help strategies? If northern Europe emerges as a winner with respect to climate change, encountering milder winters and more precipitation, will it gain from sharing this windfall with developing countries or will it hunker down into a more defensive and containment-based approach to security?

Two trends are possible. The first is that there will be a strong pull towards isolationism and self-reliance. The second is that there will be a pull in the opposite direction, towards positive-sum international cooperation. Will climate security accord more to a scenario of 'rivalry' or 'cooperation'? International relations could go either way; climate change raises the stakes in both directions. These are not mutually exclusive absolutes. In practice there will inevitably be a balance in governments' mix of responses, moving along a continuum from coordinated international action, to defensive adaptation, to aggressive competition.[39] The pertinent question is where on this continuum the EU positions itself, with which combination of cooperative and realist dynamics.

One side of the argument is that climate security will require precisely the kinds of international cooperation, rule and holistic concept of security that have distinguished the EU's self-identity in foreign policy. John Ashton and Tom Burke argued early on that all issues related to climate change must be treated as an integral part of foreign policy and that, 'The notion of a Common Foreign and Security Policy makes no sense unless it includes a clear strategy on climate as a foreign and security policy issue.' If it did respond to the challenge, they predicted that the EU should have a comparative advantage in this policy domain due to its wide array of policy instruments dovetailing closely with the multi-faceted nature of the climate security agenda.[40]

Climate security is perhaps the ultimate international public good, and the issue with most potential to be a 'cooperation multiplier' not just a threat multiplier. Some academics argue that the cooperative approach is likely to gain further prominence to overcome problems of collective action and the free-riding of states currently outside the main Western-oriented forums.[41] David Held sees climate change as the vanguard issue in a nascent cosmopolitanism.[42] Influential futurologist James Martin sees cooperative institutional structures as the key to preparing for future uncertainty: as we cannot be certain of the future effects of climate change, policy needs to focus on setting up the rules and protocols for managing whatever impacts do emerge.[43]

Climate change is often said to be the best example of a domestic–international cross-over issue, and thus not subject to the 'black box' statist logic of traditional realism. Climate insecurity will place an even higher dividend on supporting economic development, as it is this that will ultimately help poorer states defend themselves. This in turn will reduce the scale of environmental migration. Dealing with climate change requires more of a focus on good governance standards in all countries. It should encourage rich countries to invest more in helping poor and medium-sized states to undertake robust climate adaptation. And as countries are no longer able to grow certain types of crops, food security will hinge around more free trade, not illusory quests for self-sufficiency. Nobel economics laureate

Elinor Ostrom offers a perspective that stands in contrast to the environmental security school, by showing how scarcity of resources might lead to cooperation and sustainable use instead of conflicts.[44] Climate experts have recently begun to assemble data designed to show how climate policies can be integral to spurring more harmonious global coordination on a whole range of security- and health-related problems.[45]

If the International Energy Agency (IEA) is right that $5 trillion needs to be mobilised by 2050 for low carbon investments, then more integrated international markets are needed to send the right price signals. Market competition across borders is most likely to find the best low carbon solutions. From this institutionalist perspective, the germane question is how international institutional structures need to be revised to adapt to changing conditions, so as to preserve and deepen cooperative interdependence, not to re-invent geopolitical realism.[46] International trade and competition will deepen in order to ensure that new green innovations are taken up around the world.[47] From this perspective, the self-help approach is simply not tenable: international cooperation will be forthcoming as a structural necessity.

Chiming with this liberal security framework, many experts predict a new climate-security governance model, involving both public and private capital working together. The rise of climate change challenges is now commonly related to the increasing prevalence of hybrid forms of governance, entailing synthesis between traditional state and international bodies, on the one hand, and trans-local networks linking cities and regions, on the other hand – forms of agenda-setting that focus on the need for inter-societal and not just inter-state security.[48] A green-active civil society will be crucial to providing information and solutions to the broader impacts of climate change, while vibrant democracies can be more responsive to the emergence of new strategic challenges. Increasingly, strategies for addressing climate change cannot be de-linked from those aimed at deepening global democracy.[49]

For other writers, in contrast, climate change is prime accelerator of a zero-sum logic returning to international politics.[50] Climate change has certainly been seen as connected to the rubric that associates successful 'securitisation' with antagonistic and exceptional counter-measures.[51] Experts see the main geo-strategic impact as flowing from global institutions losing their credibility and being increasingly ignored as they are unable to find their way through the divisive politics of a global deal on emissions.[52] Critical analysts insist that zero-sum rivalry is built into the very way in which militaries have pushed climate security onto the policy agenda: a form of securitisation, they argue, that effectively precludes positive cooperation with and from non-Western states.[53]

Clive Ponting has chronicled how environmental stress and depletion have engendered conflict throughout history. Examples include fighting associated with the reduction in fertile land at the onset of the last ice age; the decline of the earliest cities in Mesopotamia; declines in China, Japan, Ethiopia, Lebanon and Greece thousands of years before Christ; the political aftermath of poor harvests resulting from lower levels of Nile flooding in ancient Egypt; the demise of the Roman empire; the fatal in-fighting of the Maya; and conflict in cold-snap seventeenth-century Europe. We may simply be returning to the historical norm of climate shifts being a source of major human conflict, after a brief interregnum in which it appeared as if man had, at least in select parts of the world, mastered nature.[54]

Notable political theorists such as Francis Fukuyama have argued the climate change is driving an incipient return to a 'Malthusian world'. In the last 250 years of post-Industrial revolution political development has been predicated on the state's distribution of benefits in a context that permitted unsurpassed levels of wealth expansion. The coming Malthusian world will entail a reversion to the politics of predation and force, in which 'increased military or administrative capacity is … a more efficient use of resources than investment in productive capacity'.[55]

A frequently made prediction is that climate change will make it harder to sustain basic openness and interdependence between countries, in terms of trade, investment and travel; economic localism will become more prevalent.[56] Targeted by environmental non-governmental organisations (NGOs), free trade could be an early casualty – climate is a readily added justification for protection when economic nationalism is already making a comeback.[57] Jeremy Rifkin predicts the end of the 'empathic society': empathy between nations has been predicated upon the very economic model that can no longer be sustained.[58]

The growth of rising powers brings with it an increasingly Darwinian competition for resources.[59] The battle for control over renewables and their associated supply routes will become as fierce as the international politics of hydrocarbons.[60] A tilt towards adaptation reflects a containment-based approach to security – in the hope of keeping climate-induced instability away – to the detriment of efforts to tackle the driving causes of climatic change.[61]

The iconic James Lovelock insists that all the talk of turning back global warming is useless and that the only option is a 'lifeboat mentality', on the part of those countries that will be less affected and where some quality of life may be sustained. In international relations, he conjectures, 'the ethics of a lifeboat world' will and must increasingly prevail. This will entail pumping up military capacities to keep out the hordes flocking from states harder hit by the ravages of global warming.[62] Several authors foresee an encroaching 'environmental authoritarianism', as climate change will undermine the conditions necessary for democracy to flourish. Resource conflicts will endanger fragile democracies, while efforts to turn back environmental migrants will encourage disrespect of international human rights norms.[63]

Conclusion

With non-Western powers rising fast and a range of new threats emerging, the international system is in flux. In this process of change there are elements of both rivalry and cooperation, of both zero- and positive-sum dynamics. No single theoretical interpretation of the reshaped world order has yet gained a fully convincing ascendancy. Many factors influence the balance between these realist and liberal dynamics. The question of relevance here is whether the climate security agenda will tip the balance one way or the other in any decisive fashion. As this chapter has demonstrated, there are well-grounded arguments on both sides of the debate.

The implication is that these competing analytical frameworks need to be tested rigorously against actual climate security policies. So far, insufficient empirical, contemporary evidence has been collected to reach any firm judgement one way or the other. Where international relations theories have been deployed in relation to climate issues it has mostly been in explanation of negotiating dynamics leading up to set-piece multilateral agreements.

Chapter two unpacked the different policy challenges associated with climate security. It broke down the general set of climate change predictions into a series of concrete policy dilemmas. This chapter adds further analytical orientation by comparing the contrasting predictions of how such dilemmas will be addressed. For each policy dilemma, the 'rivalry' and 'competition' framework would expect different policy outcomes. This is not to suggest that EU policy is likely to follow one or the other of these approaches exclusively; the challenge is to determine what *combination* of these rather schematic realist and liberal dynamics are adopted by different actors within the EU's *sui generis* system of security governance. The contrasting outcomes can be summarised as follows.

The Rivalry Framework

Policy challenge: To prepare for climate-related military missions and conflict prevention.

Expected outcome: EU prioritises additional investment in defensive military capability to repel spill-over effects of climate conflict.

Policy challenge: More structured diplomacy and incentives focused on climate-related global dynamics.

Expected outcome: Increased tension with strategic partners over climate security differences and less EU effort to make concessions in the name of cooperative security frameworks.

Policy challenge: Economic and development policies broadened to temper resource scarcity.

Expected outcome: The EU develops a hard-headed form of geo-economics to protect its material interests in response to scarce resources and the shift to low carbon commerce.

Policy challenge: Deal with and pre-empt new patterns of international alliances that pivot around climate threats.

Expected outcome: The EU selects its partners with climate goals more prominently in mind and employs tough realpolitik against those states seen as problematic from a climate security perspective.

Policy challenge: To mitigate stresses on democracy by moving away from state-centred approaches to security.

Expected outcome: Not strictly applicable, but in the spirit of the EU responding to populist pressure for greater isolation, the protection of national borders and reduction in migration.

Policy challenge: To deal with unpredictability through more agile response mechanisms at all levels.

Expected outcome: The EU relies more on self-help independence as the means of tempering uncertainty.

The Cooperation Framework

Policy challenge: To prepare for climate-related military missions and conflict prevention.

Expected outcome: The EU increases resources invested in comprehensive approaches to conflict prevention, with select and internally approved military intervention in cases of serious conflict.

Policy challenge: More structured diplomacy and incentives focused on climate-related global dynamics.

Expected outcome: The EU incorporates incentives related to climate security into the broad range of its civilian instruments and seeks to deepen multilateral cooperation.

Policy challenge: Economic and development policies broadened to temper resource scarcity.

Expected outcome: The EU increases climate-related development cooperation and implements a more rules-based set of international economic policies.

Policy challenge: To deal with and pre-empt new patterns of international alliances that pivot around climate threats.

Expected outcome: The EU widens its network of cooperative security partnerships through prioritisation of climate security challenges.

Policy challenge: To mitigate stresses on democracy by moving away from state-centred approaches to security.

Expected outcome: The EU develops a more participatory system of security governance and increases its efforts to support democracy and human rights around the world in ways more tightly linked to the management of climate stress.

Policy challenge: To deal with unpredictability through more agile response mechanisms at all levels.

Expected outcome: The EU pursues transparent multilateralism as a means of tempering uncertainty.

A third option is a form of null hypothesis, namely that climate change will entice EU foreign and security policies in neither one direction nor the other, but remain largely imperceptible in its impact on policy outcomes. This chapter has also shown how the EU's broader security identity might lead us to expect more of a tilt towards the liberal–cooperative framework. However, it also points out that this security identity is still contested and evolving; it may not be firmly enough rooted for us to assume that the Union will naturally eschew rivalry and realpolitik. The essential point is that more systematic and empirically rooted analysis is still required much more specifically of how governments are modifying their security and geo-economic policies in response to climate concerns. We need to test whether the EU's traditional liberal–multilateralism is sufficient to infuse its climate security policies or if, in contrast, such an identity is compromised in the face of the climate challenge. It is to such an examination of EU policies that we now turn.

Notes

1 M. Telò, 'Introduction', in M. Telò (ed), *The European Union and Global Governance*, London, Routledge, 2009.
2 European Security Strategy, *A Strong Europe in a Safer World*, Brussels, European Union, 2003.
3 J. Habermas and J. Derrida, 'February 15, or what binds Europe together: A plea for a common European foreign policy', *Constellations*, 10/3, 2003, 291–97.
4 K. Smith, *European Union Foreign Policy in a Changing World*, 2nd edition, Cambridge, Polity, 2008; J. McCormick, *The European Superpower*, Basingstoke, Macmillan, 2007; M. Tèlo, *Europe: A Civilian Power? European Union, Global Governance, World Order*, Basingstoke, Palgrave, 2007; S. Biscop and J. Andersson (eds), *The EU and the European Security Strategy: Forging a Global Europe*, London, Routledge, 2008; I. Manners, 'The normative ethics of the European Union', *International Affairs*, 84/1, 2008, 45–60; S. Lucarelli and I. Manners (eds), *Values and Principles in European Union Foreign Policy*, London, Routledge, 2006.
5 For overviews of the application to CFSP of these different theoretical schools, T. Christiansen and B. Tonra (eds), *Rethinking European Union Foreign Policy*, Manchester,

Manchester University Press, 2004; Michael E. Smith, 'Toward a theory of EU foreign policy making: Multi-level governance, domestic politics, and national adaptation to Europe's common foreign and security policy', *Journal of European Public Policy*, 11, August 2004, 740–45.

6 M. Telo, Reinventing Europe blog, ecfr.eu, 23 February 2012.

7 J. Eriksson, 'Survival of the fittest, not the strongest: Why the EU will prevail in the global power game', Swedish Institute for International Affairs brief 17, 2012.

8 For a summary of EU responses to the crisis, see R. Youngs, *The Uncertain Legacy of Crisis: European Foreign Policy Faces the Future*, forthcoming.

9 P. Vennesson, 'Competing visions for the European Union grant strategy', *European Foreign Affairs Review*, 15/1, 2010, 65.

10 R. Youngs, *The EU's Role in Global Politics: A Retreat From Liberal Internationalism*, London, Routledge, 2010.

11 Mercury Project, *Organised Multilateralism: The EU in Multilateral Fora*, policy brief 4, 2012.

12 J. Krause and N. Ronzitti (eds), *The EU, the UN and Collective Security: Making Multilateralism Effective*, London, Routledge, 2012.

13 M. Smith, 'Beyond the comfort zone: Internal crisis and challenges in the EU's response to rising powers', *International Affairs*, 89/3, 2013, 653–71.

14 R. Youngs, *The Uncertain Legacy of Crisis: European Foreign Policy Faces the Future*, Carnegie Endowment for International Peace, Washington DC, Carnegie 2014, chapter 4.

15 G. Grevi and T. Renard (eds), *Partners in Crisis: EU Strategic Partnerships and the Global Downturn*, Madrid, Fride, 2012.

16 S. Keukelaire and H. Bruyninckx, 'The European Union, the Brics and the emerging new world order', in C. Hill and M. Smith (eds), *International Relations and the European Union*, 2nd edition, Oxford, Oxford University Press, 2012, p. 392.

17 F. Attinà, 'EU and the structure of government of the global political system: The case of multilateral security', paper delivered to the 22nd IPSA conference, Madrid, July 2012.

18 T. Wright, 'What if Europe fails?', *Washington Quarterly*, 35/3, 2012, 23–41.

19 A. Cottey, *Security in 21st century Europe*, 2nd edition, Palgrave, 2012, chapter ten.

20 L. Tsoukalis, 'The JCMS annual lecture: The shattering of illusions – And what next?', *Journal of Common Market Studies Annual Review*, 2011, 49, 19–44.

21 L. van Middelaar, *Le passage à l'Europe: Histoire d'un commencement*, Paris, Éditions Gallimard, 2012, pp. 11–13.

22 U. Krotz, R. Maher, D. McCourt, A. Glencross, N. Ripsman, M. Sheetz, J. Haine and S. Rosato, 'Debating the sources and prospects of European integration', *International Security*, 37/1, 2012, 178–99.

23 Future of Europe Group, *Final Report of the Foreign Ministers of Austria, Belgium, Denmark, France, Italy, Germany, Luxembourg, the Netherlands, Poland, Portugal, and Spain*, 17 September, 2012, www.statewatch.org/news/2012/sep/eu-future-of-europe-report.pdf.

24 J. Fischer, 'Provincial Europe', *Project Syndicate*, 31 October 2012.

25 C. Kupchan, 'Centrifugal Europe', *Survival*, 54/1, February 2012, 111–18, p. 111 and p. 115.

26 S. Rosato, 'Europe's troubles: Power politics and the state of the European project', *International Security*, 35/4, 2011, 45–86.

27 S. Lehne, 'The Big Three in EU foreign policy', Carnegie working paper, 2012.

28 J. Rogers, 'A new geography of European power?', Brussels, Egmont paper 42, 2011.

29 D. Stokes and R. Whitman, 'Transatlantic triage? European and UK "grand strategy" after the US rebalance to Asia', *International Affairs*, 89/5, 2013, 1087–1107.

30 D. Keohane, 'Fresh start or more of the same? Defence policy and the December EU summit, FRIDE policy brief', December 2013, p. 2.
31 A. Marrone, 'Defence spending in Europe in light of the economic crisis', Rome, Istituto Affari Internazionali working paper 12/27, October 2012.
32 European Council, Conclusions on security and defence cooperation, 19/20 December 2013, p. 10; *EU Observer*, 19 December 2013.
33 M. Clarke, 'Rethinking Security and Power', in D. Held and D. Mepham (eds), *Progressive Foreign Policy*, Cambridge, Polity, 2007, p. 28.
34 C. Meyer, *The Quest for a European Strategic Culture: Changing Norms on Security and Defence in the European Union*, Basingstoke, Macmillan, 2006.
35 European Action Service, *European External Action Service Review 2013*, Brussels, EAS, p. 5.
36 Keohane, 'Fresh start', p. 3.
37 F. Ledwidge, *Losing Small Wars*, New Haven, Conn., Yale University Press, 2011.
38 J. Lindley-French, 'Strategic pretence or strategic defence?', GCSP policy paper 14, April 2011, p. 4.
39 N. Mabey, 'What does the security community need from a global climate regime?', E3G memo, November 2009.
40 J. Ashton and T. Burke, 'The geopolitics of climate change', *SWP Comments*, 5, May 2004, 2 and 3.
41 A. Goldthau, 'The public policy dimension of energy security', in B. Sovacool, *The Routledge Handbook of Energy Security*, London, Routledge, 2011, p. 133 and p. 142.
42 D. Held, *Cosmopolitanism: Ideals and Realities*, Cambridge, Polity, 2010, especially p. 170 and p. 223.
43 J. Martin, *The Meaning of the 21st Century*, London, Eden Project Books, 2006, p. 281.
44 A. Tor, 'The Sahel and the climate security debate', in OECD, *Global Security Risks and West Africa – Development Challenges*, Paris, OECD, February 2012.
45 D.L. McCollum, V. Krey, K. Riahi, P. Kolp, A. Grubler, M. Makowski and N. Nakicenovic, 'Climate policies can help resolve energy security and air pollution challenges', *Climatic Change*, 119/2, 2013, 479–94.
46 A. Goldthau and J. Witte, 'The role of rules and institutions in global energy: an introduction', in A. Goldthau and J. Witte (eds), *Global Energy Governance: The New Rules of the Game*, Washington, Brookings, 2010.
47 M. Kahn, *Climatopolis*, New York, Basic Books, 2010.
48 P. Khanna, *How to Run the World: Charting a Course to the Next Renaissance*, New York, Random House, 2011; N. Dubash and L. Rajamani, 'Beyond Copenhagen: Next Steps', *Climate Policy*, 10, 2010, 593–99, p. 596; German Advisory Council on Global Change, *World in Transition: A Social Contract for a Great Transformation*, Berlin, German Advisory Council on Global Change, 2011.
49 P. Burnell, *Climate Change and Democratisation: A Complex Relationship*, Berlin, Heinrich Boll Stiftung, 2009; P. Burnell, 'Democracy, democratization and climate change', *Democratization*, 19/5, 2012, 820; A. Giddens, *The Politics of Climate Change*, Cambridge, Polity, 2009, p. 158.
50 R. Beardsworth, *Cosmopolitanism and International Relations Theory*, Cambridge, Polity, 2011, p. 12; G. Rachman, *Zero-Sum World: Politics, Power and Prosperity after the Crash*, London, Atlantic Books, 2010.
51 B. Buzan, O. Waever and J. de Wilde, *Security: A New Framework for Analysis*, Boulder, CO, Lynne Reinner, 1998.
52 A.Sentance and R. Betts, 'International dimensions of climate change', *Climate Policy*, 12/1SUP:01, 2012, 1–5.

53 S. Dalby, *Security and Environmental Change*, Cambridge, Polity Press, 2009.
54 C. Ponting, *A New Green History of the World*, London, Penguin, 2007.
55 F. Fukuyama, *The Origins of Political Order*, New York, Ferrar, Straus and Giroux, 2011, p. 464.
56 . A. Kaletsky, *Capitalism 4.0: The Birth of a New Economy*, London, Bloomsbury, 2011, p. 317; T. Friedman, *Hot, Flat and Crowded*, London, Penguin, 2009 edition; N. Mabey, *Facing the Climate Security Threat: Why the Security Community Needs a 'Whole-of-Government' Response to Global Climate Change*, GMF policy brief, November 2010, p. 2.
57 P. Driesser, *Eco-imperialism: Green Power, Black Death*, Washington, Free Enterprise Press, 2003.
58 J. Rifkin, *The Third Industrial Revolution*, London, Palgrave, 2011.
59 R. Sharma, *Breakout Nations*, New York, Norton, 2013 edition, p. 225.
60 A. Behrens, 'The role of renewables in the interaction between climate change policy and energy security in Europe', *RELP*, 1, 2010.
61 B. Lomborg, 'Time for Europe to show real vision on climate change', *Financial Times*, 10 December 2010, p. 13; B. Lomborg, *The Sceptical Environmentalist*, Cambridge, Cambridge University Press, 2001.
62 J. Lovelock, *The Vanishing Face of Gaia: A Final Warning*, London, Penguin, 2009, p. 12 and p. 62.
63 M. Beeson, The coming of environmental authoritarianism', *Environmental Politics*, 19/2, 2010, 276–79; D. Sherman and J. Smith, *The Climate Change Challenge and the Failure of Democracy*, Westport, CT, Praeger, 2007.

4

THE EU'S CLIMATE SECURITY RESPONSE

Concerns over the geopolitical effects of climate change have existed for some time. Gradually, the European Union (EU) has begun to react, both through common institutions in Brussels and member states' national policy initiatives. This chapter outlines the nature of the EU's basic commitments in the area of climate security. An impressive range of formal policies has been introduced and the EU has mapped out in comprehensive fashion the degree of change required of European security policies. Nearly all member states, as well as the External Action Service (EAS) and European Commission, now have supposedly high priority policies on climate security. Ministerial speeches have honed in on the strategic implications of climate change with increasing frequency and have promised extensive action.

While a significant breadth of policy innovation is evident, much debilitating institutional inertia remains. Institutional processes have not been developed that accord climate security clear lines of responsibility. This has militated against any significant momentum being generated behind the practical implementation of climate security strategies. Notwithstanding European efforts to have the United Nations Security Council (UNSC) assume a role in climate security, in general the EU has struggled to advance holistic notions of security at the multilateral level. There is some evidence that the importance attached to climate security has begun to weaken since 2010–11, in some measure due to the deflecting influence of the eurozone crisis. Overall, the EU can be said to have established the foundations for a meaningful commitment to climate security, but has been slow to equip its institutions for significant advances at the level of policy implementation. Civil society organisations and businesses have begun to buy into the climate security agenda; both these sets of actors retain a degree of ambivalence, however, that has militated against a fully multi-actor model of EU security governance.

General commitments

The seminal moment of the EU acknowledging the geopolitical dimensions of climate change came in 2007. In that year, the EU initiated an internal reflection on climate security that engendered several policy commitments. In 2008 the then foreign policy high representative, Javier Solana, published a joint paper with the European Commission, *Climate Change and International Security*. This recognised climate change as a 'threat multiplier' that needed to be placed at the heart of EU security policy. It warned that the risks were not just humanitarian but political and strategic, affecting the EU's own interests. The paper provided a comprehensive outline of the challenges: intensified conflict over resources; more and costly humanitarian operations; new disputes over land and maritime borders; increased environmental migration and pressure for migrants to be granted refugee status; more state fragility, accompanied by radicalisation; more pressure for nuclear energy, with problems for non-proliferation; tensions between north and south, but also between rising powers and least developed countries. Challenges would vary: in Africa, fragility; in the Middle East, water wars; in Asia, coastal mega-cities beset by new vulnerabilities. The paper noted that the effects of climate insecurity were already evident, not a matter of future speculation. The paper served primarily to put down a marker and flag up the need for more systematic policy designs. Its own recommendations were not particularly decisive: it called mainly for 'more work to be done' and dialogue on climate-security impacts to be included in the EU's various regional partnerships. It claimed rather breezily that the issue played to the EU's strengths, as it required holistic responses – this was phrased almost as if impending climate apocalypse were a boon for the EU.[1]

After the paper was released, follow-up took the shape of discussions on the topic with global partners, efforts to move climate security on to the UNSC's remit and a series of seminars and commissioned studies. On the back of the 2008 report, the Commission and Council Secretariat set up an 'Informal steering group on climate change and international security'. Germany, Sweden, Denmark, France and the UK have played prominent roles in keeping this forum active in Brussels. This represented a more formalised version of the Green Diplomacy Network (GDN) that had started functioning in 2003, with the aim of feeding environmental factors into foreign policy deliberations. In fact, the EAS re-launched and also took over chairing the GDN in 2012, specifically with the aim of beefing up its hitherto low profile and mobilising it as a resource to mainstream foreign policy questions within climate change deliberations.

The Commission and Council have organised several events and consultations on climate security both in Brussels and within the context of the UNFCCC. The EAS has created a series of climate/related thematic networks that incorporate geographical departments. Diplomats highlight that regular training for officials has been introduced on the security impacts of climate change and that a plethora of studies has been commissioned. Some academics detect a closely knit 'epistemic community' of EAS, Commission and member state officials that now concur on the need to push climate security much higher on the policy agenda.[2]

The first European Union Climate Change Programme covering the years 2000–04 contained nothing on security impacts. The second programme, beginning in 2005, did mention political issues relating to adaptation, but the list of funded projects did not reveal anything security-related.[3] These flagship EU climate change programmes still lacked any kind of geopolitical flavour or tone. Recognition crept into EU pronouncements that this situation needed correcting. In the 2008 revision of the European Security Strategy, climate change was identified as a core strategic and not merely environmental challenge. The new document expanded at length on the importance of understanding climate change as a security issue – a whole section being dedicated to this topic, compared with just a few cursory sentences in the original 2003 version.[4] In March 2010 a European Council summit was for the first time devoted to discussing the strategic priorities for the international dimensions of energy policy, and linked this discussion with climate issues.[5] Governments commissioned work on climate security scenarios in 2010, and concluded with commitments to carve out a role for the largest regional security organisations focused on early prevention and to delve into regional specificities.[6] Officials highlight that the EU has begun to map member states' vulnerabilities to water insecurity, and done this with EAS involvement.

The Lisbon Treaty gave the EU formal competence in energy policy. While the treaty did not introduce any new provisions specifically of relevance to the geopolitics of climate change, diplomats saw its new energy clause as a useful foundation for articulating a tighter climate–security linkage.[7] The general effort to strengthen EU foreign and security policy has begun to permeate climate policy. Climate action commissioner Connie Hedegaard suggested that the 2011 Tunisian revolution was related to climate change, to the extent that this lay behind the rise in food prices that triggered social protests against the Bin Ali regime; a climate strand was therefore to form a part of the EU's response to the Arab Spring.[8]

In July 2011 the EU agreed new council conclusions on 'climate diplomacy'.[9] Officials admitted that by 2011 much momentum had been lost from the climate security agenda and that the statement was a 'new starting point'. The council conclusions were triggered by an Anglo-German letter complaining that the EAS needed to begin engaging in a more tangible and systematic fashion on the foreign policy dimensions of climate change; a climate unit that had originally existed in the Relex department of the European Commission had actually been broken up and been transferred out of the foreign policy orbit since promulgation of the seminal 2008 paper. The new climate diplomacy strategy promised identifiable action on the security strand of EU policy.

In September 2011, the Commission produced a long-awaited memorandum on how the EU should develop a more strategic approach to its international energy policies. The role of climate change policies within the EU's core security policies was a central thread to the new policy document. The new strategy reinforced the commitment to linking climate change policies to the EU's core security policies. A number of proposals more specifically pertinent to climate change were also advocated, in particular to enhance cooperation with international partners on renewable energy.[10] The document's tone was very much that effective strategies against climate change could be an opportunity for the EU to regain strategic influence and leadership.

In July 2012 the External Action Service released an upgraded policy document on the Arctic. In this, the EU committed to strengthening its security role in the Arctic region. The document stated: 'Maintaining good international cooperation in the Arctic region and supporting the region's stability is a key interest of the European Union.' It stressed the importance of a holistic approach based on the nostrums of 'knowledge, responsibility and engagement'. Additional resources were to be forthcoming for sustainable development in the region, especially in support of indigenous communities. A €1 million development impact assessment of climate change was to be carried out. The EU was to press more assertively its bid to gain observer status in the Arctic Council, pending since 2008. Diplomatic engagement with Arctic states would be intensified. Bilateral EU–Greenland relations would be prioritised, in correction of their strikingly under-developed profile to date. The EU would hinge its engagement on ensuring the United Nations Convention on the Law of the Sea (UNCLOS) develop a more effective role as a route to rules-based cooperation across the Arctic. The EU promised to contribute to partnership-based shipping rules and emerging procedures.[11] Notwithstanding this, in May 2013 the Arctic Council again postponed an offer of observer status to the European Union, while granting this to China and five other states (Germany, France, the UK, the Netherlands, Poland and Spain already have observer membership). In late 2013 a number of northern states sought to attract EU aid for projects in the Arctic, extending the focus of cross-border cooperation under the Union's so-called northern dimension, citing geopolitical changes as justification.[12]

A European Parliament foreign affairs committee motion of October 2012 served to refocus debate on the issue within the Brussels institutions, and called on the EAS and Commission to forward more specific policy commitments in the area of climate security.[13] Also in October, the Commission proposed to ring-fence 20 per cent of the 2014–19 EU budget for climate questions across all policy areas; this would include external relations resources being deployed far more meaningfully for the geopolitical dimensions of climate change. The European Council agreed this target in February 2013.[14] The EAS began working with DevCo, the Commission's development agency, to decide how to take forward this commitment, with the aim of injecting some degree of political focus into the way that the new funds were to be deployed.

At the end of June 2013 the foreign affairs council adopted conclusions on climate diplomacy and security, together with a new External Action Service reflection paper. The fact that the issue of climate diplomacy was included in a foreign affairs council packed with so many immediate emergency issues was said to reflect the importance attached now by foreign ministers to this longer-term agenda. The conclusions and reflection paper together provided a new high-level political impetus and a number of operational commitments. The foreign ministers' conclusions boldly opine that effective climate action is a 'necessary condition for peace and security'. They commit ministers to an annual review of progress made in injecting foreign and security policy parameters into climate change strategies. The reflection paper claims that the EU is now much more 'joined-up' in having its mainstream diplomacy dovetail with climate objectives, especially through a reinvigorated GDN.[15] Another ministerial in July 2013 focused on water challenges, with the aim of enhancing EU engagement in the Mekong, Nile and other worsening water-related tensions.[16] Some of the momentum from 2008/2010 appeared to be returning, through these and other new policy documents on adaptation, geo-economics and climate, and climate migration, all detailed in subsequent chapters. At this stage, the EU also stepped up efforts to promote a new UN discussion on the topic, set for September 2014, with the backing of the UN secretary general.

The June 2013 council conclusions committed the EU to mainstream climate security in all key external dialogues. In April 2013 the Group of Eight (G8) adopted a new focus on climate security.[17] Climate security dialogue deepened in particular within EU–US relations from 2012. European diplomats undertook an extensive outreach campaign in 2013 in the US 'heartland' as well as through Washington events designed to explore how to incorporate climate change into transatlantic security cooperation. The EU and South Africa inched forward with a first climate dialogue that explicitly incorporated security questions. The EU has since then gradually begun to inject a more political edge into its various bilateral climate dialogues. Insiders insist this is particularly the case with Brazil and China. The China dialogue was raised to ministerial level in 2010 and given a €500 million boost at the end of 2011. A new sustainable urbanisation initiative between the EU and China kicked off in 2012.[18] Chinese officials insist they are now eager for cooperation on climate issues as they recognise this will be the factor that most limits China's growth. Significantly, the EU has set up a regional platform for cross-border water management in Central Asia. The EU's December 2012 communication on Maghreb cooperation also promised to bring North Africa into the reach of EU energy and climate rules and to work with the region on jointly addressing 'climate-induced risks'.[19] A number of EU-backed initiatives have sought to establish international dialogue and research agendas related to the broader impacts of climate change. For example, the SE4ALL initiative aims at an international network to strengthen 'resilience', while the Euroclima initiative fosters dialogue on the social impact of climate change in Latin America. Several European leaders made a link between climate change and the November 2013 typhoon Haiyan in the Philippines.

Member states follow suit

Developments at the EU level have been complemented by those within member states' national policies. The German foreign ministry has stated that 'policy must be active in more than just the [UNFCCC] negotiating process', and that hence the government was developing 'climate and security' as a priority 'new area of activity'.[20] Germany has pushed for the development of a 'geopolitical change management that takes into account the effects of changing physical environments on governance structures'.[21] The German minister of state called for 'an effective brand of preventive climate diplomacy through close coordination in a spirit of trust and partnership'; climate security should become a core topic around which broader international alliances are constructed in order to head off security tensions, in exactly the same way that proliferation or human rights have been recognised as mainstream strategic issues.[22] The German federal foreign office began increasing its climate change capacity from 2010. Funds for additional posts and a series of practical initiates were forthcoming after 2011 to help follow-up on Germany's climate security agenda in the UNSC. Foreign ministry diplomats in the climate diplomacy unit work to sensitise other departments to the issues related to climate security with a view to mainstreaming the latter across all other areas of policy. The German federal foreign office and the UK Foreign Office have begun systematic cooperation on climate security.

German policy has been especially focused on the UN; the country's 2011 efforts to embed a climate security remit in the UNSC are portrayed as its main contribution to this agenda. Germany launched a programme of initiatives after its 2011 UNSC chairmanship to generate momentum behind the broader geopolitical dimension, with the stated aim of getting foreign ministries in other countries to think more in terms of climate security. This involved an extensive series of regional initiatives in, respectively, Africa, Asia, Latin America, Central Asia and the Middle East. These aimed to elaborate proposals on how to confront climate change from a security perspective using regional dialogues to foster south–south networks and impart best-practice experiences. In September 2012 foreign minister Guido Westerwelle hosted an informal event on climate security with his Moroccan counterpart in the United Nations. Building on this, the UK and Pakistan joined forces for a follow-up informal debate in early 2013. Germany has engaged Norway on a joint seminar programme, reflecting concern that the latter is heading towards the Canadian and Russian line of unrestrained resource exploitation in the Arctic. The federal government has used a share of revenues from the ETS specifically to fund research and educational programmes on the geopolitical impact of climate change.

As early as 2003, then British prime minister Tony Blair warned of the dangers of climate change to security, insisting that 'The destruction of the environment and global warming are as great a threat to world peace as terrorism and weapons of mass destruction.'[23] The UK's 2008 *National Security Strategy* posited a link between climate change and security and claimed that 'Climate change is potentially the greatest challenge to global stability and security, and therefore to national security.' A priority focus was the danger of other states stockpiling resources in the face of shortages.[24] A Department for Environment, Food & Rural Affairs (Defra) official leading national adaptation plan insists that the approach has been to 'inject the international dimension, because otherwise we could make no progress on things like national floods, which is all people think [climate security] is to do with'. Foreign secretary William Hague's climate envoy has repeatedly insisted that climate security policies only made sense as a common and cooperative endeavour at the global level.

Officials report that climate change already impacts upon planning and policy analysis in Afghanistan and African conflict interventions. The UK defines climate change as a critical driver of such civil conflict threats. A climate group was formed and began feeding in assessments to the UK National Security Council. The UK set up a parliamentary committee on climate change, also with a broad mandate covering the more political dimensions of global warming. In 2010 William Hague suggested that 'Climate change is perhaps the twenty-first century's biggest foreign policy challenge'. Responding effectively to all the different dimensions of climate change is above all else a job, he argued, for foreign policy: diplomacy had the responsibility to 'put climate change at the heart of foreign policy'.[25] In July 2011, the UK Foreign Office published its *Building Stability Overseas Strategy*. The document refers to climate as a source of instability multiple times. The document states that the government 'will further encourage the [United Nations Security] Council to mainstream climate security into its conflict work' and also commission 'systematic reviews of the evidence … on the impact of climate change'.[26]

At the end of 2011 the UK government produced one of the most comprehensive reports to date disaggregating the relevance of climate change across each major area of foreign policy. This report formed part of the UK government's new legal commitment to offer a more political set of climate risk assessments. It pointed to the government's attempt to assess how climate change threatened not only direct physical effects but would bring 'new interests into the geopolitical calculations of states, for which current global governance structures were not designed'. The prevailing institutions of global governance were designed to prevent war between ideological adversaries, not to manage shared problems of resource scarcity. This government analysis pointed strongly in the direction of a cooperation-based approach to climate security; indeed, it argued that the gravest climate security threat could be its undermining of multilateral cooperation even more than its direct effects on the domestic environment.[27]

Indeed, along with Germany, the UK has been one of the most active of European states in pushing to have the UN Security Council assume a lead role in climate change. In 2007 then foreign secretary Margaret Beckett argued that climate change was, 'an issue which threatens the peace and security of the whole planet – [the UN Security Council] has to be the right place to debate it'.[28] The UK Ministry of Defence (MoD) has acknowledged that the challenge of climate security increases the need for cooperative multilateralism, stating that:

> There is clearly a role for MoD to play in the understanding of, and response to, the international impacts of climate change; however this cannot be achieved by acting alone. Rather, a combination of actors is necessary in order to both avert the likelihood of climate change induced conflict, and to respond if and when it does occur.[29]

Denmark was also an early promoter of the linkage between climate change and security. In a 2006 report, the Danish Ministry of Foreign Affairs recommended that Denmark 'use its influence in the UN system to try to get anthropogenic climate change recognised as a threat to international peace and security'.[30] Denmark is the only country that is a member of the EU, NATO and the Arctic Council. Its ministry of foreign affairs has an all-encompassing department for environment, energy and climate change. In 2009, Denmark hosted a high-level seminar on climate change and international security. The then foreign minister, Per Stig Møller, argued:

> Climate change, if left unchecked, will intensify and multiply existing security threats. Security threats like demographic pressures, access to water,

> and food shortages. The alarming thing is that it will be the most vulnerable regions that will face the most severe consequences of climate change.[31]

Denmark was possibly the first country to make concrete calls for the whole array of international organisations to recognise climate change as a security threat.

Sweden's general commitment to tackling climate change is habitually ranked amongst the highest in the world. Sweden has created an inter-ministerial working group with participants from the prime minister's office and ministries of environment, defence and foreign affairs that commits itself to addressing climate security policy issues. Crucially, Sweden sees the Arctic Council as a primary tool for addressing the geopolitical dimensions of climate change, and as a good example of effective multilateralism. Swedish diplomats routinely claim that the Arctic Council has made an unprecedented degree of cooperation possible among the traditionally mutually suspicious Arctic states. With some irony, it is often pointed out that the high level of commitment flows from these two Scandinavian countries that are widely seen as potential winners from global warming. It is said to be testament to the fact that even those states less directly ravaged by climate change now acknowledge that they will be menaced by damage done to other parts of the planet.

In 2010, the Netherlands played a lead role in galvanising debates over food security and the move towards 'climate smart agriculture'. The Dutch government committed to additional climate funding for 2013 directed in particular at international cooperative forums working on contested water management. France has expressed particular concern with the effects of emissions in Africa and most specifically with the implications these could have for security in the Sahel region. In 2009, it launched a joint Anglo-French study on the security implications of climate change in the Sahel region. The study aimed to determine the security implications of climate change for the period from 2030 to 2040.[32] President Sarkozy proposed a World Environmental Organisation that would have a broad political mandate and serve as focal point for more joined-up multilateral commitments. Building on these commitments, the French government published an extensive climate strategy in 2011 that also broadened out the geo-strategic focus of climate action.[33]

Spain's 2011 national security strategy talks of climate as representing a particularly acute threat, given the country's geographical position and proximity to North Africa.[34] In 2005 the Spanish government set up a Military Emergencies Unit (Unidad Militar de Emergencias) to respond to natural catastrophes, including those caused by climate factors. Climate change features as one of five global issues for the Italian ministry of foreign affairs, alongside human rights, the fight against terrorism, disarmament, and energy[35] – despite former prime minister Silvio Berlusconi being a prominent climate change doubter; at the 2008 G8 summit he famously suggested that prioritising climate issues in the midst of an economic crisis was 'like someone with pneumonia thinking about having a hairdo'.

The proactive role adopted by a number of member states in pushing for UNSC coverage of climate security has brought European governments into conflict with rising powers and developing states that prefer to see climate change dealt with in the UN's development and social bodies, and not treated institutionally as a primarily security-related topic. This became a hotly debated issue during Germany's 2011 UNSC presidency. Two years on from his original, flagship speech on climate security, Ban Ki-moon spoke again on the climate change–international security nexus, at a July 2011 UN debate on the topic. The secretary general argued that the geopolitical link had become even more palpable; predicted that 'mega crises may well become the new normal'; and lamented that there was still a need for 'joining the dots' between climate change policy and different areas of security policy.[36] The UN debate in 2011 was seen as successful to the extent that China did not object to the UNSC at least discussing the issue. But in this debate, Russia, China and most developing nations rejected the European argument that the Security Council should have a formal mandate to address the security implications of climate change. Nauru's permanent representative to the UN reacted angrily, arguing that if the UN did not appoint a special representative for climate change and security then action would be too tardy and limited; she reassured that such a figure would not cut across the UNFCCC and would not lead to a spate of blue helmet military interventions designed to restrict countries' use of energy – both fears expressed by opponents of a UNSC mandate.[37]

European support provided a fillip to these initiatives. European efforts at the UN constitute an important pillar of the EU's climate security strategy, even if progress here has been limited. The UN secretary general has asked for help from the EU in preparing several of his papers and speeches on climate security. Senior diplomats insist that the EU has tried to raise climate security questions at the Copenhagen, Cancun, Durban, Doha and Warsaw summits, from 2009 to 2013, even if it was rebuffed by other powers in doing so.

Institutional inertia

This plethora of commitments, at member state and EU level, represents genuine and significant progress. Climate security is now firmly on the EU agenda. Climate change is conceived as an issue relevant to security strategy in a way that was not the case a decade ago. However, the EU's foreign policy machinery – in Brussels and the national capitals – has still declined to accord climate security an unequivocal priority. The familiar problems of institutional coordination that beset EU external policies in general are particularly acute in climate security, as this issue adds a further level of internal policy competences to the various strands of foreign and security policy deliberation.

Limitations are reflected in the relative paucity of concrete follow-up on the EU's seminal 2008 paper and other formal climate-security commitments. Many policy-makers acknowledge in private that core aspects of the debate have stagnated at the level of conjecture. Diplomats recognise this is a serious issue but remain frozen in terms of knowing what to do within their security as opposed to environmental policies. Nick Mabey summarises in sober fashion: 'There is a mismatch between analysis of the severity of climate security threats and the political, diplomatic, policy and financial effort being expended to avoid these risks.'[38] Experts working closely with European government initiatives acknowledge that 'transferring efforts for policy learning to translate existing knowledge and legal frameworks into decision-making is still in its infancy'.[39]

It is striking that after a period of intensive activity, few significant policy commitments were forthcoming at the EU level in 2011 or 2012, with the exception of the paper on the Arctic region. As said, a number of new documents produced in mid-2013 re-injected some degree of momentum; however, some diplomats still lamented what they perceived as an overall diminution in the pace of policy initiative compared with preceding years. Climate security was hidden away as an apparently low-profile issue on the newly operational website of the External Action Service. An extensive search of member states and EU sources reveals few policy documents or initiatives directly related to climate security that were forthcoming in 2012 or 2013. The aforementioned 2012 American Security Project's *Climate Security Report* mostly repeated well-known challenges but also succeeded in galvanising debate on a new 'non-state centric' security 'rethink' among the Washington security establishment. Similar momentum was less evident in Brussels.[40]

Strategists have begun to reflect on the startling absence of a 'European geostrategy' aimed at thinking through how the EU should be securing access to global trade routes, to temper the dangers of strategic bottlenecks and deal with climate-security linchpin states around the world.[41] Critics decry that none of the profound changes that will occur to global political geography are anywhere near being integrated into strategic planning.[42] One of the abiding weaknesses of CSDP is that it functions in a 'strategic void', especially in failing to match conflict interventions with strategic interests emanating from patterns of oil and gas management and climate change.[43] One leading European diplomat opines in private that the geopolitics of climate change is the one area of 'new threats' where the EU has failed 'to move beyond the theory'. Experts and policy-makers alike admit that while the UK was one of the first states to adopt the most holistic, forward-looking discourse on merging climate and foreign policy in the mid-2000s, tangible change in its security policies is hard to identify. A German diplomat laments that the debate is 'not as vivid now' as attention has been refocused to the economic crisis and the Arab Spring.

Policy-makers ritually acknowledge the lack of institutional momentum. EAS staff lamented that the 2012 Rio Summit did not attempt to address climate security and that nobody in the EU machinery pushed for it to do so. Privately, insiders lament that the experience of the 2009 Copenhagen summit – at which the EU appeared to be excluded from key deal-making involving the US and rising powers – was to make the EU more cautious in future COP meetings and to make it more focused on simply 'being in the photograph' rather than pushing the debate into considering more comprehensive political linkages. When the EAS launched a new energy and foreign policy paper in August 2012, it contained nothing on climate security.

EU diplomats lament that while CSDP personnel have been 'sensitised' to the issue, there is still no high-level 'champion' to link the EU's security arm to climate security. No equivalent of the UK's climate security envoy has been appointed at the EU level. High representative Catherine Ashton has not given a single speech dedicated specifically to the issue. One EAS diplomat acknowledged that as of 2013 the EU was still failing to move beyond the traditional CFSP agenda fully to deal with new challenges such as climate security. Notwithstanding the significance of the June 2013 council conclusions and reflection paper, these talk primarily about using foreign policy to push other countries to raise their levels of 'climate ambition' for a post-2015 UN agreement on emissions. On the more strictly defined security aspects of policy, the commitments in these documents are still phrased mainly in terms of 'developing a narrative', 'awareness-raising' and 'engaging' with business, civil society, other powers and international organisations – that is, they are largely about the EU's own institutional preparedness and couched less in terms of commitments to tangible policy output.[44]

The EU has failed to streamline institutional structures. Overall capacity in climate issues decreased in the move from Relex to the EAS. Responsibilities for foreign and security policy, energy security and climate change are split in confusing fashion amongst a large number of institutional players. In Brussels, after many years of internal debate, the precise division of competences between the External Action Service, DG Energy and DG Climate Action remains imprecise. None of these departments is tasked unequivocally with leading on the geopolitical impact of climate change. A 2013 European Parliament report lamented that energy policy was the area where coordination remained poorest between the External Action Service and the Commission, and by extension between the foreign policy dimension and more technical areas of external cooperation and funding.[45] Institutional links have remained especially weak between DG Environment and the other departmental players – DG Energy, the External Action Service and DG

Climate. Climate security has become one of the clearest examples of an issue that falls into the gaps between ministerial portfolios and institutional mandates.

Remarkably, after producing the seminal 2008 paper, the CFSP machinery discontinued its main climate security contact position. A specified climate security post in the new EAS was only formed in late 2011. While EU documents claim that the holistic nature of climate insecurity represents an opportunity, the requirement for even more effective coherence also reveals one of the EU's abiding weaknesses. The EU and its member states still do not have a list of climate security risks that slot into the foreign policy machinery with a specified ranking of priorities to guide decision-making. There is much talk and frequent speeches on how any number of ills will be magnified by global warming; but there is no apparent listing of which current and actual high security risks are indeed climate-related. There is still no single institutional source taking charge of a grand strategy for climate security. Moreover, environmental and security imperatives are still debated nationally, not primarily at a European level and even less so at a wider multilateral level.

Ministers and officials opine in private that climate security still has no resonance as an issue with the public. It is widely felt that a committed epistemic community has emerged on climate security but that this still struggles to capture high-level political attention. Many of those working on the ground, especially in developmental areas, still tend to associate 'climate security' with 'militarisation' and thus recoil from the agenda. It is also feared that talking in terms of security could endanger smooth relations with non-European countries. DG Climate and member states' environment ministries insist that they have kept security issues out of the UNFCCC process, as they fear this would be a distraction from emissions targets. Experts note that where coordination has improved it has been in tightly delineated sectors of environmental policy, not on the wider nexus between climate issues and more essential political questions. The GDN in particular has failed to make the jump from such sectoral concerns to gain any traction in pursuit of a more politicised approach to climate security.[46]

Indeed, some member states – the French government being the most vocal – have expressed dissatisfaction with this situation and pushed the EAS to assume more of a balanced, shared leadership with DG Energy on the external dimensions of energy policy. In consultations for the 'EU2020' energy strategy, member states complained strongly that the EAS remains woefully under-equipped in energy capacity and expertise. Companies like Statoil expressed concerns about their operations suffering from the institutional 'muddle' between the EAS and DG Energy. One Dutch diplomat catches a broader concern, opining that, 'all actors are trying to institutionalise the agenda in a traditional sense rather than develop qualitatively new approaches to the governance of climate security'. Some officials now insist that the EU is too unwieldy and that smaller groupings of states offer the only hope of advancing climate security strategy.

Coordination problems persist at the national level too. In Germany differences between the different parties forming the governing coalition have held back the climate security agenda; prior to the September 2013 federal elections, the FDP led through the German federal foreign office, while CDU and CSU ministries remained more ambivalent. In some member states, climate issues have palpably slid down the list of priorities. Spain's main foreign policy strategy from 2011 failed to make any mention of climate as a foreign policy challenge.[47] The annual review of Spain's foreign and security policy in 2012 made no mention of international climate policy.[48] By the end of 2012, climate security was absent from the list of Spanish security thematic priorities.[49] The new PP government that took office at the end of 2011 was not strongly committed; its leaders sometimes openly question the priority attached to climate change.[50]

European states organise their coverage of climate change in very different ways. France and the UK combine climate and energy policy; in the UK within their own joint ministry, in France through jointly operated responsibilities of the ministries of ecology and of finance. In Italy and in Spain, climate change departments answer to overarching environment policy; in Italy the issue is dealt with in a department that also includes renewable energy and sustainable development, which in Spain are dealt with by the ministry of industry. In Germany, climate change depends on the ministry of the environment, but its division also amalgamates international environmental cooperation. Sweden and Poland have both climate change and international divisions within the ministries of the environment, but in a less consolidated fashion than in Germany. Of particular relevance here, it is striking to note that climate change's inter-linkages with institutions covering finance, industry, energy, economy, transport and other aspects of environmental policies are rarely matched by formal connections with foreign or defence ministries.

There is a common feature across member states' different institutional set-ups: the traditional community of security policy-makers remains relatively marginal. Most member states' foreign ministries still have a limited involvement in climate security; this partly explains Europe's limited influence over the way that climate security is addressed by other powers around the world.[51] Environment ministries may have adopted the discourse of 'geopolitical impact' but they have fought hard to keep the security community at arm's length. There remains a pressing need to broaden decision-making out beyond energy and environmental ministries, who have the incentive to present their own control of policy in this area as adequate.[52] For example, DG Energy officials admit in private that they have sought to avoid the EAS gaining a prominent role in energy, as they fear that the foreign policy lead would then be too strong and cut across the EU's basic energy model predicated on the extension of internal market rules and linkages. Traditional energy experts and policy-makers still exhibit a tendency to dismiss foreign policy aims as extraneous distractions. In DG Energy many bemoan the accords signed for foreign policy reasons with key supplier countries that, they say, now stand hollow.

The UK is one of the few European countries where climate change features in the title of a top-level ministry. The UK's department for energy and climate change was created in 2008, but is generally seen to have done little to integrate its two component parts, energy security and climate change. The UK has made much progress in ensuring that different departments and units from across Whitehall meet regularly to share reflections on climate security. However, while the UK climate change board was given a climate security mandate, it has considered the issue rarely and in very general terms.

A wider set of deliberations on how climate security goals impact on the EU's broader set of global alliances appears even further away from materialising. Notwithstanding a small number of more developed 'climate dialogues', there is little evidence that climate security has become a significant factor conditioning the shape of EU global alliances and strategies for effective multilateralism. The UK and other European states are charged with failing to adjust their international alliances in accordance with how climate change is likely to rebalance global power.[53] One study finds that the EU has been unable to use its bilateral partnerships with rising powers to entice the latter to sign up to multilateral climate deals or to make linkages with broader politically related topics.[54] In private, participants describe the political aspects of EU energy dialogues with strategic partners as 'empty'.

So, for example, diplomats acknowledge that the climate partnership with China has struggled to gain momentum, with Beijing ambivalent over the CCS initiative, the urbanisation project and EU support regionally for the Mekong river initiative; moreover, they fear that a new China–US climate dialogue risks excluding the European geo-strategic influence over climate security. The EU (and US) have also failed to adapt their geo-strategic alliances to the emergent importance of Atlantic Basin energy dynamics; it is argued that this will represent an important geopolitical framework conditioning the shift to low carbon, to which EU states have paid insufficient heed as they remain focused on traditional energy alliances with Russia and Middle Eastern hydrocarbon producers.[55] Some insiders lament that in practice the EU has backed away from its previous position that the Arctic should be managed under a regime of shared cooperative security; it has come to accept the notion of parcels of territory being carved out between the different states, as envisioned by Russia in particular. The EU's hoped-for influence over Arctic management has thus not prospered.

Officials in Brussels and the member states admit that the focus has been almost entirely on restructuring internal institutional structures rather than developing a comprehensive multilateralisation of climate security. An Institute for Environmental Security (IES) report found that the 'security dimension of climate change is not as widely covered in bilateral policy dialogues with third countries as it could be'.[56] European diplomats lament that there remains a need for one overarching body to coordinate the broader effects of climate change. They have advocated some kind of joining together of OPEC, the IEA, the International Renewable Energy Agency and the Major Economies Forum (renamed from the Major Emitters Forum in 2009). In practice, little progress has been made. European states have not invested any great effort in prioritising this issue within their multilateral diplomacy. Such oversights belie the EU's commitment to a broad, multilateral, cooperation-based prioritisation of climate security.

Participative security governance?

What of the assumption that climate security pushes in the direction of more multi-actor clusters of security governance? On the broad issue of climate change, a strikingly large number of initiatives has proliferated involving non-governmental organisations (NGOs), companies, municipal authorities, consumer groups and individual citizens, some local and some transnational. Arguably, this has injected greater momentum than anything achieved through UN negotiations. Some observers argue that this range of non-state activity now extends beyond an ad hoc collection of initiatives into an increasingly organised alternative structure of climate governance.[57] The US National Intelligence Council and EU Institute for Security Studies concur that climate change is an issue that has begun to engender looser forms of governance based on target setting, peer pressure and informal codes of conduct. Over 300 public–private partnerships have been created since the 2002 Sustainable Development conference, providing a foundation from which these types of venture can begin to influence the more geopolitical dimensions of climate change.[58] The United Nations Development Programme (UNDP) reports that the picture is no longer one of purely northern civic activism in the environmental field; rather, south–south links between advocacy NGOs have come increasingly to influence the global climate agenda and advance the notion of 'ecological citizenship'.[59]

Many member states certainly claim to be in favour of a 'new diplomacy' that links together governments, international bodies, social networks, entrepreneurs, NGOs, knowledge institutions, advocacy and interest-based groups. Incipient coordination in Brussels between environmental, development and human rights NGOs has broadened the perspectives taken by the longest-standing climate change lobbies.[60] In its range of proposals forwarded in September 2011 for a new external energy and climate change policy, the Commission explicitly urged the systematic involvement of industry and energy companies.[61]

Some organisations like the IES have adopted the agenda and mobilised to raise its profile. An IES event in the European Parliament at the end of 2012, for example, focused on the fact that member states and the EAS had declined to follow the UK's example by appointing envoys for climate security.[62] The innovative Earth Security Initiative has played a similar role. Its experts lament that after the original hype of climate security, progress has stalled as it has become clearer that climate security means very different things to different stakeholders and it remains unclear how to give concrete substance to the agenda. Climate security remains an abstract and 'fluffy' agenda as governments have failed to convince electorates and parliaments that the issue of resource scarcity will be integral to very tangible national interests in the future. Northern European countries remain sanguine as they perceive they will be relatively unaffected; in many NGOs' view, this has sapped the spirit of intra-EU solidarity. An agenda of 'climate conflict' must make way for one of 'climate peace'. This needs to move beyond concerns with 'greening the military'. Climate security should give way to a broader focus on strategies to address resource scarcity. To make progress the agenda needs to be presented as one of economic opportunity not a threat to be contained. Earth Security Initiative experts lament that little has been done to back multilateral regional fora such as those set up to manage cross-border water resources in the Nile Basin and Himalayas. Governments have failed to develop a positive message that the climate security agenda could help improve the earth's resilience and that it is not a matter of holding-containment.

Such forward-thinking organisations have made an impact. However, civil society organisations themselves remain divided over the climate security agenda. Core bodies like the Citizens (London) Forum and Sustainable Energy (Bucharest) Forum created to engage civil society on efficiency questions have not embraced high politics security questions. The excellent and highly influential European Climate Foundation has also eschewed an explicit security focus in its targeting of the global climate agenda.[63] A trawl through Friends of the Earth documents, press releases and statements reveals how the organisation studiously avoids the standard lexicon of climate security; rather, it approaches the theme extremely obliquely, focusing on the 'human cost' of climate change, food security and health problems, an assertive campaign against EU support for biofuels in developing states, critiques of the insufficiency of the EU's new energy efficiency rules and 'environmental injustice'.[64] NGOs like Saferworld that have pushed hard for conflict-reducing initiatives to be built into adaptation programmes are criticised by development organisations. NGOs campaigning on the 'resource curse' in developing states have pushed the EU to prioritise transparency measures in the hydrocarbon sector but have remained ambivalent if not hostile to the incipient diversion of resources to renewable projects in these economies.[65] The notion of citizens being able to exert 'climate justice' through litigation has not progressed, and is not a notion backed by governments; this has led some experts to recall that a global UN-level agreement is still necessary and cannot be entirely replaced entirely by bottom-up approaches.[66] Environmental NGOs are also deeply split on geo-engineering.[67]

Companies have been even more low-profile and tepid in their views on the climate security agenda. Of course, a plethora of business organisations have been formed that cover climate issues, including the World Business Council for Sustainable Development and the Corporate Leaders Group on Climate Change. But, little ethos of public–private partnership has taken shape on climate change's more geo-strategic dimensions. Business Europe – the main employers' organisation in Brussels – has remained disengaged. Its engagement on the broader questions of climate change has been described by insiders as bland and humdrum. The European Roundtable has been sceptical of attempts to link foreign to environmental policy. Renewables companies tend to be small and have struggled to access formal, strategic-level decision-making processes.[68] The Energy Business Council was launched in March 2009 to ensure that the voice of the energy sector is heard in the EU policy debate. The weight of its members – which include the likes of Acciona, Alstom Power, Chevron, Edison, Enel, Eni, Eurelectric, Fortum, Iberdrola, Petrobras, Renault, Repsol, RWE, Shell, Siemens and Statoil – militates in favour of a focus on very traditional 'security' questions. A Shell representative observed that interest in climate security has diminished since the economic crisis, with the company now lobbied less from NGOs and solicited less frequently by governments on this question.

Statoil executives lament that little progress has been made in changing the nature of the policy process. Much more systematic input is still required from companies into the design of EU external energy and climate policy. Only ad hoc and superficial changes have been introduced so far. Few links have been built up between the private sector and the Network of Energy Security Observers.[69] The 2010 survey of the Investor Network for Climate Change found growing preparation for climate change issues among investors, but with limitations: a lack of knowledge persisted among investor analysts of climate change linkages to standard investment indicators, and less than a third of investors were found to have engaged directly with policy-makers on the broader climate change agenda.[70] Climate change is the issue that investors have found hardest to factor into political risk and where most uncertainty exists of all the issues routinely looked at by companies, weakening their policy engagement.[71] Many companies remain strikingly recalcitrant. Nearly half of executives admit that the broader implications of climate change to their global operations are not built into their corporate planning processes.[72] Many multinational corporations still lobby that the climate security agenda is a fiction. The UK's 2011 Foresight report laments the lack of business involvement in security planning and risk assessments associated with climate change.[73]

Conclusion

At a formal level, the EU cannot be accused of having under-reacted to the potential security impact of climate change. A huge number of European policy documents and speeches have proclaimed that climate change will become not simply another security issue alongside other threats, but the matter that conditions all other international strategic questions. Incremental accretions to European security strategies have progressively reinforced the centrality of this nostrum. The cooperative–multilateral dimension has occupied a central place in EU thinking of how to approach climate security.

Despite all this, however, policy-makers on their own admission appear uncertain how to proceed further. The EU is renowned for its institutional 'messiness'; the climate security agenda has become strongly affected by this and entrapped within opaque lines of responsibility. Indeed, the issue's almost uniquely multivariate nature means that it has been particularly hard to establish any single institutional site unequivocally to take charge of taking forward a climate security strategy. Climate security is judged to be important in the long run but apparently not immediately urgent enough to prompt a quick improvement in the EU's institutional preparedness. It has been argued that the EU's external climate policy has been the product of domestic factors, inter-institutional rivalries and bureaucratic stickiness, much more than of an adequate response to changing global conditions and alliances.[74] Conversations with diplomats reveal a widespread feeling that while climate change needs to be understood as a security challenge, for the moment the linkage remains too abstract to elicit serious attention; it has to some extent been displaced by more immediate challenges like the Arab Spring and the eurozone crisis. It is agreed that institutional structures will accord greater priority to climate-security only when this is seen to require immediate and concrete responses in particular crisis scenarios. This may be an issue that requires long-term planning and pre-emptive action, but in practice no one institutional site is likely to seize full responsibility until a singular crisis makes the climate–security link undeniably palpable. A trigger is likely to be required in the form of a disaster somewhere where there is a clear security dimension inextricably intertwined with the effects of extreme climate.

Notes

1 EU High Representative and the European Commission, *Climate Change and International Security*, Brussels, Council of the European Union, March 2008.
2 K. Zwolski and C. Kaunert, 'The EU and climate security: A case of successful norm entrepreneurship?' *European Security*, 20/1, 2011, 21–43.
3 European Commission, *European Climate Change Programme ECCP II*, Brussels, 2005, see http://ec.europa.eu/clima/policies/eccp/index_en.htm.
4 European Union, *Report on the Implementation of the European Security Strategy – Producing Security in a Changing World*, Brussels, European Union, 2008.
5 Council Secretariat, *European Council Conclusions, 25–26 March 2010*, www.consilium.europa.eu/ueDocs/cms_Data/docs/pressData/en/ec/113591.pdf.
6 A. Maas, *Shifting Bases, Shifting Perils*, Berlin, Adelphi Research, 2010.
7 Treaty of Lisbon amending the Treaty on European Union and the Treaty establishing the European Community, signed at Lisbon, 13 December 2007.
8 Presentation at the European Business Summit, May 2011.
9 Council Secretariat, *Council Conclusions on Climate Diplomacy, July 2011*, http://ec.europa.eu/clima/events/0052/council_conclusions_en.pdf.
10 European Commission, *On Security of Supply and International Cooperation – The EU Energy Policy: Engaging with Partners beyond our Borders*, COM (2011) 539.
11 European External Action Service, *Developing an EU Policy towards the Arctic Region: Progress Since 2008 and Next Steps*, JOIN (2012) 19, direct quote from p. 11.
12 *EUobserver*, 'Arctic region to see greater focus in EU aid', 9 October 2013.
13 European Parliament Foreign Affairs Committee, *The Role of the CSDP in Case of Climate-driven Crises and Natural Disasters*, European Parliament report, October 2012.
14 Council Secretariat, *Council Conclusions, 7/8 February 2013*, www.consilium.europa.eu/uedocs/cms_data/docs/pressdata/en/ec/135344.pdf, p. 6.
15 Council Secretariat, *Council Conclusions on EU Climate Diplomacy*, foreign affairs council meeting, Luxembourg, 24 June 2013; External Action Service, *EU Climate Diplomacy for 2015 and Beyond: Reflection Paper*, Brussels, 2013.
16 *Council Conclusions, July 2013*, press release, www.consilium.europa.eu/uedocs/cms_Data/docs/pressdata/EN/foraff/138317.pdf.
17 www.gov.uk/government/news/g8-foreign-ministers-meeting-statement.
18 European External Action Service, *EU-China Sustainable Urbanisation Partnership*, http://eeas.europa.eu/delegations/china/eu_china/sustainable_urbanisation/sustainable_urbanisation.htm.
19 European Commission and High Representative, *Supporting Closer Cooperation and Regional Integration in the Maghreb*, JOIN (2012) 36, December 2012, p. 15.
20 German Federal Foreign Office, *Climate Change: A Foreign Policy Challenge*, www.auswaertiges-amt.de/EU/Aussenpolitik/GlobaleFragen/Klima/Aussenpolitische-Dimensione-node:html.
21 German Federal Foreign Office, *Climate Diplomacy: Reducing Risks for Security*, Berlin, 2012, p. 17.
22 C. Pieper, 'Climate diplomacy in perspective', in D. Tänzler and A. Carius (eds), *Climate Diplomacy in Perspective, from Early Warning to Early Action*, Berlin, German Federal Foreign Office, 2012, p. 20.
23 www.guardian.co.uk/politics/2003/feb/25/uk.environment2.
24 Her Majesty's Government, *A Strong Britain in an Age of Uncertainty: The National Security Strategy*, London, The Stationery Office, 2008, p. 18 and p. 27.
25 W. Hague, 'The diplomacy of climate change', Speech to the Council on Foreign Relations, 27 September 2010.

26 www.fco.gov.uk/resources/en/pdf/publications/annual-reports/bsos-july-11.
27 Foresight, *International Dimensions of Climate Change*, London, UK Government Office for Science, 2011, p. 9 and p. 40.
28 M. Beckett, UN Security Council debate – 2007, www.guardian.co.uk/environment/2007/apr/18/-.
29 UK Ministry of Defence, *Defence in a Changing Climate*, London, The Stationery Office, 2010.
30 Danish Ministry of Foreign Affairs, *Climate Change and Foreign Policy – An Exploration of Options for Greater Integration*, Copenhagen, Danish Foreign Office, 2006.
31 S. Møller, opening speech of the Seminar on Climate Change and International Security in Copenhagen, 2009.
32 P. Heinrig, *Security Implications of Climate Change in the Sahel Region: Policy Considerations*, Paris, OECD, 2010, www.oecd.org/swac/publications/47234320.pdf.
33 Gouvernement de la France, *L'action exterieur de la France contre le changement climatique*, Paris, Gouvernement de la France, 2011.
34 Gobierno de España, *Estrategia Española de Seguridad: Una Responsabilidad de Todos*, Madrid, Gobierno de España, 2011, p. 37.
35 www.esteri.it/MAE/EN/Politica_Estera/Temi_Globali/.
36 Secretary-General's remarks to the Security Council on the Impact of Climate Change on International Peace and Security, 20 July 2011.
37 M. Moses, 'Needed: A UN special representative on climate change and security', *Huffington Post*, www.huffingtonpost.com, 28 July 2011.
38 N. Mabey, 'Facing the climate security threat: Why the security community needs a 'whole-of-government' response to global climate change', GMF policy brief, November 2010, p. 4.
39 D. Tänzler and A. Carius, 'Overview: Towards preventive climate diplomacy', in D. Tänzler and A. Carius (eds), *Climate Diplomacy in Perspective, from Early Warning to Early Action*, Berlin, German Federal Foreign Office, 2012, p. 20.
40 C. Foley and A. Holland, *Climate Security Report*, Washington DC, American Security Project, November 2012.
41 European Geostrategy at http://ideasoneurope.eu.
42 C. Paskal, 'From constants to variables: How environmental change alters the geopolitical and geo-economic equation', *International Affairs*, 85/6, 2009, 1143–56.
43 S. Biscop and J. Colemont, *A Strategy for CSDP: Europe's Ambitions as a Global Security Provider*, Egmont, October 2010.
44 Council Conclusions, *Council Conclusions on EU Climate Diplomacy*; European External Action Service, *EU Climate Diplomacy*, p. 5.
45 European Parliament, *The Organisation and Functioning of the European External Action Service: Achievements, Challenges and Opportunities*, Brussels, European Parliament, February 2013, p. 83.
46 B. Lee and D. Torney, 'Leading on climate and resources', in G. Grevi and D. Keohane (eds), *Challenges for European Foreign Policy in 2013: Renewing the EU's Role in the World*, Madrid, Fride, 2013, pp. 36–37.
47 A. Baron, 'El cambio climatico y la acción exterior de España: Un necesario cambio de perspectiva', *Fride blog*, 18 December 2012, www.fride.org.
48 Ministerio de Asuntos Exteriores y Cooperación, *2012: Un Año de Legislatura en Política Exterior*, Madrid, Ministerio de Asuntos Exteriores y Cooperación, 2012.
49 www.maec.ec, Seguridad Internacional.
50 www.elpais.com/articulo/sociedad/Rajoy/cuestiona/cambio/climatico/afirma/puede/convertirse/gran/problema/mundial/elpepusoc/20071022elpepusoc_7/Tes.

51 D. Torney, 'European climate diplomacy: Building capacity for external action', Finnish Institute for International Affairs briefing paper 141, October 2013.
52 Mabey, 'Facing the climate security threat', p. 6.
53 D. Depledge and T. Feakin, 'Climate change and international institutions: Implications for security', *Climate Policy*, 12/1 SUP01, 2012, 73–84, p. 82.
54 G. Grevi and T. Renard (eds), *Hot Issues, Cold Shoulders, Lukewarm Partners: EU Strategic Partnerships and Climate Change*, Madrid, Fride, 2012.
55 P. Isbell, *Energy and the Atlantic: The Shifting Energy Landscape of the Atlantic Basin*, Washington DC, German Marshall Fund, 2012.
56 G. de Ville and R. Kingham, *Recent Trends in European Union External Action in the Fields of Climate, Environment, Development and Security*, The Hague, Institute for Environmental Security, 2011, p. 34.
57 M. Hoffmann, *Climate Governance at the Crossroads*, Oxford, Oxford University Press, 2011.
58 EU Institute for Security Studies-National Intelligence Council, *Global Governance 2025: At a Critical Juncture*, Paris, EU Institute for Security Studies, 2010.
59 United Nations Development Programme, *Human Development Report 2013*, Geneva, UNDP, p. 111.
60 R. Wurzel and J. Connelly, 'Environmental NGOs: Taking a lead?', in R. Wurzel and J. Connelly (eds), *The European Union as a Leader in International Climate Change Politics*, London, Routledge, 2011, p. 216.
61 European Commission, *On Security of Supply*, p. 13.
62 www.envirosecurity.org/events/201211.php.
63 See www.europeanclimate.org.
64 See www.foe.uk.
65 L. Fuhr and S. Wykes, 'Perspectives on resource politics in a climate constrained world: How "resource curse" activists view climate change', *Democratization*, 19/5, 2012.
66 R. Lord, J. Brunnée, S. Goldberg and L. Rajamani, *Liability for Climate Change Damage*, Cambridge, Cambridge University Press, 2012; G. Fülbermayr, 'Climate change liability: Transnational law and practices', *Climate Policy*, 13/1, 2013, 145–47.
67 International Institute for Strategic Studies, *Strategic Comments*, 18/46, 12 December 2012, 3.
68 W. Grant, 'Business: The elephant in the room?', in Wurzel and Connelly, *The European Union as a Leader*, pp. 202–3 and 205.
69 See eeas.europa.eu/energy/network_en.htm.
70 Investor Network on Climate Change, *Global Investor Survey on Climate Change*, 2010, p. 7 and p. 23.
71 I. Bremmer and P. Keat, *The Fat Tail: The Power of Political Knowledge for Strategic Investing*, Oxford University Press, 2009, p. 10.
72 R. Sullivan, 'An assessment of the climate change policies and performance of large European companies', *Climate Policy*, 10, 2010, 38–50, p. 40.
73 Foresight, *International Dimensions*.
74 S. Schunz, 'Explaining the evolution of European Union foreign climate policy: A case of bounded adaptiveness', *European Integration* online papers, 16/ 6, 2012, p. 19.

5

EU CLIMATE AND ENERGY POLICIES

Foundation or diversion?

In examining European Union (EU) approaches to climate security, it is necessary to consider the Union's well established and distinctive role in both general climate policy and traditional energy security. This presents a challenge that derives from the EU's unique set of energy rules and regulations. The Union's policies are predicated strongly on a deeply embedded range of regulatory activity. As pointed out in chapter three, this is invariably seen as forming the basis for much European influence beyond the EU's formal borders. This is not akin to traditional security policy, but has an implied strategic dynamic. It is impossible to conceptualise EU 'actorness' in climate security without taking this on board. The question is what implication EU climate and energy security policies have for its contribution to climate security. Do the Union's mainstream climate and energy policies spill over to accord it an effective and distinctive identity in climate security? Or is the link between these policies tenuous and even negative for climate security?

This is not a book about the EU's climate change policies per se or the organisation's performance in multilateral climate negotiations. However, to set the scene for an assessment of global warming's impact on EU security policy it is necessary to provide an overview of the state of play in Europe's overarching climate policies. Experts and European officials themselves often claim that the EU's lead role in climate policies enables it to influence the broader political dimensions of global warming. But are the EU's climate commitments really exemplary enough to lay the foundations for it to play a lead role in dealing with these geopolitical challenges? Likewise, the chapter assesses the relationship between the EU's traditional energy security policies and climate security aims. The Union has prioritised efforts to develop a more unified and effective 'external energy policy' in recent years. It has made much progress in doing so. In this nascent external energy policy, a comprehensive definition of climate security is said to be integral to traditional security of supply concerns. But, has the EU's energy security policy really given such a positive prompt to climate security? With so much focus now on the potential development of shale gas, does the climate security agenda still have the same meaning and resonance? In relation to both these ostensible linkages the chapter finds as many grounds for arguing that longer-standing, core EU policies – climate and energy, respectively – detract from as much as they contribute to effective strategies for climate security.

A climate change leader?

In many respects, the EU is justified in claiming to be a climate change leader. Germanwatch's 2012 climate change performance index placed six EU member states – in order, Sweden, the UK, Germany, Denmark, France and Belgium – in its top ten most committed countries worldwide.[1] Progress has been made on each of the EU's flagship '20/20/20' commitments: renewables, emissions and efficiency. The IEA has reported that most member states are on track to meet their target to have 20 per cent of their energy generated from renewables by 2020; two-thirds of new generating capacity in the EU now comes from renewable sources; and most of the growth in renewable energy still comes from Europe, even during the current economic crisis.[2] On emissions, the more ambitious target of securing 80 per cent reductions by 2050 has come to dominate policy deliberations. The EU agreed a new energy efficiency directive in June 2012, which came into force in December 2012. It sets efficiency targets within public procurement rules. The EU has the lowest 'energy intensity' of all regions (measured as energy supply per unit of gross domestic product (GDP)) and the highest demand for renewable energy.[3]

Low carbon technology now represents a €300 billion market and provides employment to over 3 million workers in Europe. The Commission has supported over a dozen large-scale pilot projects on carbon capture and storage (CCS). The 2009 EU Energy Programme for Recovery committed €4 billion of investment in infrastructure and interconnections, alongside renewable projects. Of this total, €1 billion went to CCS projects. A first licence for the commercial implementation of a CCS project was granted in France in 2011. In 2010, the European Investment Bank (EIB) channelled a record €19 billion of credits to low carbon initiatives, a 20 per cent increase from 2009 and two-thirds of all EIB loans in Europe. Across all its various budget lines, by 2011 the Commission was putting over €1 billion into 'frontier' low carbon research and development.[4]

The Danish government that took office in November 2011 made a commitment to pursue the goal of having the country's entire electricity and heat supply come from renewable sources by 2035. In 2010 the UK launched a Green Investment Bank, with £3 billion of capital. For the period 2011–14, the German government increased its research and development funding for green technologies by 75 per cent over the preceding three-year period, partly in response to the decision to phase out nuclear power generation by 2022.[5] Late in 2011 the EU agreed to make €300 million of Emissions Trading Scheme (ETS) revenues available for CCS and other renewables projects through the EIB. In April 2012 the British government launched a new £1 billion scheme for the commercialisation of CCS, and another £125 million for research on CCS.[6]

At the UNFCCC Durban summit in December 2011, the EU pushed hard and on the basis of its proposed accord China, India and the US finally agreed to emissions targets with legal force – albeit only from 2020. And 35 states agreed to a second round of post-Kyoto treaty commitments. At the follow up Doha summit in December 2012, the EU stressed the positive from what most interpreted as an underwhelming summit, insisting it had kept doubters on board with preliminary work towards a new international agreement from 2015.

These all represent significant advances. However, the EU's general performance on climate change policy has been far from faultless. Many observers doubt the logic and impact of the EU's talismanic '20/20/20 by 2020' strategy: its arbitrariness is clearly reflected in the convenience of all the numbers being 20. The EU decided against moving unilaterally from the 20 to a 30 per cent emissions reduction target. The effect of the economic crisis is such that the 20 per cent emissions target is now achievable without great effort; even the 30 per cent target would not require much additional reform. Moving from 20 to 30 per cent reductions would only cost the EU 0.1 per cent of GDP, according to the Commission. Yet despite this, the EU has promised to increase to a 30 per cent reduction target only if others follow suit. Moreover, this was not a firm or ambitious enough commitment to make states like the US calculate that they would be better off in terms of net welfare gain by increasing their own offers.[7] Emissions targets may be met by 2015, yet several member states refuse to contemplate any revision that would enjoin meaningful structural change prior to 2020. After the EU agreed to suspend a tax on non-EU airlines' emissions for one year in 2013, member states backed away from supporting firm action after the suspension lapsed.

DG Energy has lamented that 'the quality of National Energy Efficiency Action Plans, developed by member states since 2008, is disappointing'.[8] Yet, while the 2012 energy efficiency directive has created a new legislative framework, member states are still to be subjected to binding targets for efficiency gains. The directive was watered down to exclude any binding obligations or control mechanisms on member states.[9] A new competitiveness document agreed at the June 2013 European Council was seen as cutting across any ambitious 2050 climate change agenda and heralding the pre-eminence of a 're-industrialisation' dynamic.[10]

In March 2013 the Commission published a green paper to begin consultations over a framework for energy and climate policies up to 2030. This hit the headlines by intimating that it might be preferable to have a binding target only for emissions in 2030, not renewables or efficiency.[11] Environmental organisations consequently slammed what they saw as a decrease in ambition on the part of the Commission.[12] The green paper acknowledged that by making its offer of a 30 per cent emissions target conditional on others' similar moves, the EU had not brought forward new pledges from around the world, and that a post-2015 international agreement remained highly uncertain.[13] Crucially, the EU has come to add more conditions on its commitment to a new post-Kyoto deal.[14] Poland in particular has blocked ambitious targets for 2050, arguing that these should be made conditional on other powers' commitments.

Some member states, like the UK, have moved to the higher 30 per cent target, in binding fashion. They argue that these moves should be Europeanised if the EU as a whole is to retain credible climate leadership. The UK has been one of the European states most committed to combating climate change but still needs to double its rate of emission reductions to meet its long-term targets. Renewables still make up only around 3 per cent of UK power, well short of the country's 2020 target. Moreover, the UK lags behind other Organisation for Economic Co-operation and Development (OECD) states in low carbon research and development which is at too limited a level to make any notable impact.[15] In autumn 2011 the British chancellor, George Osborne, caused waves when he declared that henceforth the UK would seek to move no faster on green commitments than its EU partners, in an effort to conserve jobs.

Respected expert Dieter Helm insists that the EU has only made progress on its emissions targets by dint of the collapse of Soviet-era industries in Eastern Europe, a switch from coal to gas and now the economic recession. Moreover, the EU has not constructed the infrastructure for feeding significant amounts of renewable energy into the grid. Governments have focused on subsidies for renewables, but the broader structure of the energy market has not changed, meaning that even current levels of green electricity sit idle, unable to get into the grid.[16] Europe's grid simply cannot absorb sufficient amounts of renewables-generated power to meet the EU's targets. Indeed, the Commission's March 2013 green paper recognised that advances in renewables risk being undermined by the need for 'massive investment in transmission and distribution grids. … to accommodate renewable energy'.[17]

Member states' different forms of support for renewables are effectively undercutting any prospect of a single market in green energy. No European government will subsidise green energy in another member state. Some analysts fear that the variety of liberalisation (or 'unbundling') options now available to national energy champions will fragment the European market even more. Renewables-generated electricity is still not traded across borders, holding back its take-off. A proposed North Sea supergrid interconnector for wind farms at sea has been held up by national protection of home markets.[18]

The recession has gradually eaten into funding for renewables. Italy announced cuts in solar power incentives. Spain has cut subsidies for solar investments, leaving many companies in severe difficulties. The UK's Green Investment Bank is curtailed by strict rules against deficit financing. Denmark has gradually reined back on its use of wind farms as these were proving to be inefficient and of intermittent use. EU officials fret that R and D on renewables has slowed dramatically due to the economic crisis.[19] The Commission's flagship research budget, FP7, gave only €2.35 billion to low carbon research out of a €50 billion total allocation. Its successor Horizon 2020 programme inks in €5.7 billion for renewable research out of a total €80 billion budget;[20] in private, officials admit this may prove insufficient.

In the Renewable Energy Attractiveness Index, most European states have fallen down the rankings and lost ground to China and US.[21] By 2012 China accounted for over 20 per cent of new investment in renewable development. While the share of low carbon power has grown in Europe, two-thirds of this is still nuclear. In private, some member state officials complain that the Commission has adopted an extremely legalistic approach to restricting large-scale wind farms, and that member states have been unable to inject the geopolitical case for an expansion of such renewables.

The EU's much-lauded ETS has not had a dramatic impact on emissions levels. The March 2013 green paper admitted that the ETS 'has not succeeded in being a major driver towards long-term low carbon investment'.[22] The ETS's first phase was extremely lax; allowances were actually greater than emissions and the carbon price plummeted. The ETS's second phase tightened up allowances, but covered sectors that still represented only 40 per cent of EU emissions. Even in its third phase, the scheme remains well short of full auctioning. Carbon offsets compromise the scheme's ostensible rationale: nearly all EU states have carbon footprints way in excess of their national reporting, because they buy the right to pollute outside Europe. Sectors excluded from the ETS still account for 50 per cent of the EU's total emissions. In a move interpreted as an interventionist turn, in 2010 the UK announced it would set a floor to the carbon price, and that this would be well beyond that of the ETS in an effort to correct the latter's ineffectiveness. Connie Hedegaard has acknowledged that the ETS has not had a major impact but has refused to cede to non-governmental organisations' (NGOs') demands for post-2012 allowances to be revoked for some sectors. The chief of executive officer (CEO) of German energy company E.ON defined the ETS as 'bust … dead'.[23]

The ETS carbon price fell to an all-time low during 2012. Hedegaard's attempts in 2013 to revitalise the ETS have floundered against resistance from within the European Parliament and some member states. The Commission floated the idea of taking millions of emissions allowances off the market to give the ETS back some bite, but this idea was subject to considerable opposition in the European Parliament and among member states.

Also pertinent, the touted nuclear renaissance is now on hold. A majority of member states were considering moving back into nuclear power by 2010. After the Fukushima disaster in April 2011 many backtracked, in particular Germany, Belgium, Italy and (non-EU) Switzerland. Germany, Spain, Poland and others have been slow to reduce state aid to the coal sector. By 2013 Europe was the second highest user of coal in the world, as the US has switched to shale gas. European states were only on target for their emissions targets because they were relying on coal-based production in China and other markets. They have reduced carbon production on their own territories but have increased carbon consumption, simply importing goods made from carbon-rich production situated in other parts of the world. In mid-2011, the Commission started infringement proceedings against nearly all member states for having failed to implement the 2009 directive on the development of CCS.[24]

What read-over to climate security?

In short, the mainstream components of EU environmental policies have advanced but are not without their serious shortcomings. While this book does not purport to study EU climate policies per se, to note the latter's broad contours is important to the extent that these condition the link with security strategies. Experts see the EU has having developed an impressive leadership role in international climate negotiations in large measure as a derivative impact of its accumulation of internal environmental competences. The very multiplicity of actors within the EU's internal decision-making processes has created incentives for strong environmental leadership. The EU is said to have exercised both structural and cognitive leadership. This leadership has been pursued as valuable in itself and also as a means of reinvigorating the broader project of European integration as other areas of cooperation have atrophied.[25] Most academic work on the EU has focused on how Union internal competences have been used as a basis for developing a role in negotiating international environmental agreements; it is widely argued that the Lisbon Treaty has tilted the balance towards Commission influence vis-à-vis the member states and given a prompt to EU global environmental leadership.[26]

Integral to such leadership, an 'externalisation' of internal climate change policies is conceived to be an integral part of the EU's foreign policy approach. The declared policy is to export the ETS globally, along with the Renewable Energy

Directive (RED).[27] The EU has linked its ETS with that of Australia, for example. It is proposed that the multilateral Energy Charter Treaty apply its rules to low carbon sources too. Some member states advocate extending the ECT's aquis comprehensively to cover renewable sources. Some officials see relevance for the Middle East and North Africa in using the 20/20/20 targets as a form of experimental governance to galvanise cooperation on climate change mitigation. Extrapolating the same 'externalisation logic', policy-makers envision extending this proactive role into the sphere of climate security.

However, it cannot be said that the EU's commitment to mitigating climate change is so strong that there is a significant or natural spill-over of climate-related considerations into security policies. The EU regularly claims that the example of its own climate leadership serves as the foundation for an internationalisation of its influence in this area of policy. But this domestic-to-external read-over is not without blemish. The advance of EU environmental policies has certainly been pervasive enough to ensure that areas of European external policies can no longer remain completely immune from climate change considerations. But neither has their progress been so exemplary so as to guarantee a natural externalisation of climate primacy across a wide panoply of security issues. External Action Service (EAS) diplomats admit that member states' growing caution of climate change in general has infected the more specific climate security agenda.

In some senses, the expansion outwards of the EU's highly regulatory approach constrains – more than it empowers – the geopolitical dimensions of climate security policy. Security deliberation tends to get crowded out by a mindset of regulatory export. Senior members of the EP's foreign affairs committee have admonished the Commission's 'introverted' focus on replicating internal market rules and consequent blindness to the international geopolitical dimensions of climate and energy.[28] DG Energy has assumed the most dynamic and proactive de facto role on the external dimension of energy policies. This compounds the tendency for external policies to be deliberated primarily as an extension of internal energy policy – as opposed to common foreign and security policy (CFSP) leading with a priori foreign policy considerations that then build energy and climate factors into such broader strategic priorities. Indeed, states like Germany, France and the UK have pushed the climate security agenda in part to wrest influence away from the Commission's bureaucratic-regulatory mindset which they fear is inimical to nuanced geo-strategy.

The read-over between environmental and external policy is more contested than it is seamlessly reinforcing. The EU's Environmental Council is openly reluctant to associate with security policy dynamics. Many Commission DGs meet in coordination groups on environmental policy; similar coordination with the EAS on climate security is conspicuously absent. One EAS official laments that the transfer of personnel and competences to DG Climate has aggravated not improved this disconnect. DG Climate has remained cautious; its officials argue that the EAS needs to 'move into gear' on climate security, as this is not primarily a concern for the EU's mainstream climate action body. DG Climate understands its foreign policy role in the relatively limited terms of simply keeping the EAS and EU in-country delegations informed of mainstream and scientific aspects of the climate change debate; its aim is mainly to make foreign policy generalists climate-aware. There has been the occasional connection between environmental and foreign policy: the best known was when the EU made a very direct geopolitical linkage between Russia signing the Kyoto agreement and European support for Russia's entry into the World Trade Organization (WTO). But these are the exceptions that reconfirm the rule of more habitual disconnect. Indeed, linkages can be negative for climate goals. The EU postponed its carbon airfare tax in 2013 because it was concerned that this was leading it towards a general international, strategic isolation.

Energy security versus climate security?

For many years the EU's rhetoric has pointed to the need to understand 'energy security' through the lens of a longer-term horizon that brings renewables and climate-related considerations into the geo-strategic foreground. Much progress has been made on rebalancing traditional energy security and climate policies. However, the risk has emerged of the EU's reinvigorated external energy security policy crowding out climate diplomacy. To some degree, this beefed-up energy security policy has been harnessed as a platform from which to build dialogues on climate security with global producers; but it mostly pivots around efforts to shore up supplies of conventional hydrocarbon supplies as well as bringing shale oil and gas into the policy equation. Traditional policies relating to the security of energy supplies have been 'securitised' to a much greater degree than climate change.[29]

Then UK minister Chris Huhne liked to argue that the 'climate security versus energy security' debate presented a false dichotomy, to the extent that climate change is likely to disrupt the supplies of oil and gas too.[30] However, in practice, the EU's most intense area of activity remains that of efforts to secure oil and gas supplies. While initiatives purporting to prepare Europe for the geopolitical effects of climate change have remained embryonic, hydrocarbon diplomacy has intensified to a far higher level. Contrary to all the rhetoric proclaiming otherwise, the EU has engaged far more assiduously in a geopolitical scramble for oil and gas than it has moved to address the security implications of climate change. Key ministries like the British Foreign and Commonwealth Office admit to being overwhelmingly focused on oil and gas when thinking about energy security.

For a long time the EU has had more of a common climate policy than an external energy security strategy. It has often been remarked that in the EU climate policy trumps energy security policy, the reverse of US policy. The *Energy 2020* strategy published in early 2011 begins by stating: 'The same collaboration and common purpose that has led to the adoption of the EU's headline energy and climate targets is not yet evident in external energy policy.'[31] The most senior of DG Energy officials lament that member states still under-estimate the potential of a truly common EU external economic dimension of energy that would entail not just an extension of internal competences but an alignment of all economic policies with energy goals. In some senses, much of the effort witnessed recently has focused on redressing this situation and finally injecting some substance into the external dimensions of EU energy security coordination.

The EU's September 2011 external energy policy template did talk of the need for 'external energy security relations' to go beyond the issue of securing supplies of hydrocarbons and to include enhanced global partnerships on climate-related issues. However, in terms of its actual policy commitments and ideas the document focuses overwhelmingly on traditional security of supply concerns. Its coverage of sustainable energy is limited mainly to promises to invest more in helping African citizens gain better access to energy. The document, apparently the flagship of all EU external climate and energy-related policies, is devoid of the standard language and issues proper to the domain of climate security. The new strategy caught most attention by proposing a new mandatory information exchange on bilateral energy accords and a provision for the Commission to negotiate new energy treaties on behalf of member states. The information exchange rules became operational in November 2012. They give the Commission nine months to assess member state agreements' compatibility with EU rules (albeit only those agreements involving governments).

The EU has signed a plethora of bilateral energy accords. After an EU–Uzbekistan memorandum of understanding was signed in 2011, all Central Asian countries had such agreements. Under an EU–Azeri deal signed in January 2011, Baku committed to the so-called 'southern corridor'; a further accord in December 2013 tied down further deals of these supplies. Azerbaijan agreed to sell Shah Deniz gas to Europe, with BP in charge of the consortium. Commission president José Manuel Barroso made what was interpreted as a particularly significant visit to Turkmenistan in early 2011, to secure supplies for the putative Nabucco pipeline. The Council talks of the need for privileged energy partnerships that go beyond the current generation of agreements. The EU offered an 'energy roadmap to 2050' to assuage Russian calls for security of demand. Observers agree that these steps reflect a more geopolitical approach towards energy security, with European suppliers reacting to uncertainty by seeking more predictable long-term contracts with producers.

A considerable slice of EU policy is focused on the question of which of several pipeline options into European markets is to be favoured. The Nordstream pipeline directly connecting Russia with German markets started pumping in November 2011. The Commission initially committed to supporting the Nabucco pipeline, but in early 2014 this project was suspended. It may fund the Trans-Caspian Pipeline, to be finished by 2018. Poland and other member states have insisted that the EU must create and fund Platforms to help all major infrastructure projects abroad. The energy NGO Counter Balance argues that the EU is, if anything, more obsessed with large oil and gas infrastructure projects now and less focused on the broader implications of low carbon than it was in the mid-2000s. This is despite the serious setbacks encountered in all such projects. The Trans-Sahara pipeline has overshot its budget by an estimated $15 billion and struggled to find sufficient investment to progress. The opening of the Medgaz pipe was continuously postponed until 2011, by which time it was running at $1 billion over budget and had to be expensively rescued by EIB loans. These pipelines represent the very opposite of the localism needed to mitigate climate change: they entail huge environmental damage, significant energy losses in transmission and deepen reliance on hydrocarbons.[32] Climate aims remain disconnected from policy deliberation of oil and gas pipelines; the EU already enjoys an incoming pipeline capacity well in excess of what its hydrocarbons consumption must be to meet its 2050 emission targets.[33]

The EU's new energy dialogues with producer states have remained narrowly couched and have not tackled the climate-security nexus in any systematic or operational fashion. Middle East producers have been particularly resistant to broadening notions of security to include climate security issues or to cooperating with the EU on mitigation (even the richest of Gulf states are still classified as developing states and thus not subject to binding emissions reductions).[34] Member states have pushed for more high-level political involvement in energy dialogues, as a means of effecting linkages to strategic priorities. But the Alliance for An Energy Efficient Economy admonishes the EU for failing to engage more systematically in meaningful dialogue with consumer states on the broader political dimensions of energy and climate policy. Areva executives argue that the EU should be pressing hard for a new United Nations (UN) agency that has a broader remit covering all energy- and climate-related questions.

Senior officials now talk enthusiastically of a new emancipation of energy policy from climate policy. Energy commissioner Gunther Öttinger has been a notable sceptic of any more politicised climate security agenda. Experts interpreted the EU's 2050 roadmap as signalling the end of the consensus that energy policy should be unequivocally led by climate policy.[35] The economic crisis and squeeze on competitiveness, combined with a new rise in oil prices, have produced a swing away from the absolute priority attached to climate policy in previous years. Long-term energy security is increasingly a matter of the 'switch to gas'. In February 2012, a mid-term report on the EU Recovery Fund revealed that most projects under way were for gas power generation; none of the eligible CCS projects had started.[36] The UK's March 2012 budget gave a tax break for oil and gas production in Shetland.

Policy-makers' main concern is now quite clearly with the advent of sizeable shale gas supplies. Prominent academics lament that the new focus on shale gas has reinforced governments' tendency to define energy security as being a matter of the 'grand strategy' calculations of hydrocarbons, putting off even further into the distance the prioritisation of similar geo-strategic planning in relation to renewables.[37] Experts opine that combining traditional and unconventional gases, the world has 300 years of supplies. The policy priority in this sense is to delink gas from oil markets, by completing the single market in gas infrastructures and linkages.

A fundamental change in energy has occurred with the US switch to shale gas: a change which leaves Europe with higher energy costs over the foreseeable future than both the US and China. More positively, as the US has gained autonomy due to shale exploration, this has driven down LNG prices, which in absolute terms is good for European buyers. Not only do many talk of the end of pipeline politics, but the switch to shale gas has given the US its lowest emissions in 20 years. Countries like Algeria claim that they have more shale gas than natural gas. Other shale basins exist in Norway, Poland, Ukraine, Turkey and East Africa. In March 2013 the UK company Centrica signed a sizeable deal to import shale gas from the US. The most significant impact will be if supplies increase enough for China to switch from coal to shale gas. The next step would be for unconventional gas to be extended into the transport sector, which is where oil dependency is still nearly absolute.

Policy-makers have breathed a sigh of relief as gas returns to being a buyers' market. This has further lessened the perceived urgency of the climate security agenda. The policy focus has seemed to shift away from the question of access to non-European renewable sources to debates over how far shale could and should be incorporated into the European energy mix. So far, Poland has been the keenest on developing shale; France has restricted fracking; the UK gave the go ahead to preliminary exploration in 2011. A European Parliament hearing in 2012 reinforced the extent to which the new focus on shale gas now cuts across the EU's international climate diplomacy.[38]

One EU official highlights that the climate directorate has pushed for this new energy panorama to be turned to the advantage of the climate security agenda, with the argument that if traditional energy security concerns now appear less acute, then there should be more policy space to prioritise the longer-term strategic impacts of climate change. These two issues can now be delinked, where they have often been conflated or at least mixed up in the same set of overarching external energy policies. However, the climate directorate has struggled to gain any purchase with this argument against other decision-making centres concerned more with the apparent near-term shale windfall. Another senior official describes debates over shale gas as a 'big distraction' from the need for the EU to assess long-term climate-security linkages.

With shale present in many stable, advanced and friendly countries, the security worries appear less acute. Some experts say this is not entirely negative for climate change aims. Industry experts even calculate that using natural gas as a 'bridging' solution would reduce the cost of meeting the EU's 20/20/20 targets relative to the huge subsidies ploughed into wind and solar. And more environmentally friendly drilling techniques are being developed for shale. However, this focus does mean that once again debates have returned to very traditional questions of the balance between hydrocarbon exploitation and environmental concerns; qualitatively new deliberations over the supposed climate–security nexus are in danger of being squeezed from this script. Long-term, it leaves Europe just as strategically vulnerable. European shale plans have, in practice, recently hit all kinds of problems, including in Poland.

The *Energy 2020* document acknowledges that the EU needs to put much more effort into relations with consumer countries.[39] The EU has certainly recognised the need to deepen consumer–producer dialogue through the IEF. Some experts insist that the IEF has to be the institutional site where fully inclusive and broad-ranging dialogue occurs in the future. Some analysts insist that effective mitigation will advance only in the IEF and better climate financing in the Group of Twenty (G20).[40] France has also pushed for a G20 role in renewable technology and curbing price volatility in energy markets. Yet European governments have appeared intent on limiting the IEF to being an extremely 'soft' institution with few formal competences. Saudi Arabia has invested considerable effort to raise the profile of the IEF to promote consumer–producer dialogue and mutters that EU representation at some of these meetings has been disappointingly limited. European governments have also been more reticent than the US to support Chinese and Indian membership of the IEA, a move seen by many as an essential step towards a genuine multilateralisation of the full range of energy and climate issues. As yet, there is little sign of the EU seeking to harness such energy security fora to advance with commitments related to the broader agenda of climate security.

Conclusion

In sum, the chapter has revealed that the implications of these well-established and mainstream areas of policy are not entirely positive for the EU's commitment to effective climate security policies. To some extent the climate security agenda has percolated into the policy consciousness as an extension of the EU's lead role on the core aspects of international climate negotiations. However, while the EU has a relatively strong record in emissions reductions and renewables uptake, its mainstream climate change policies are not so impressively far-reaching that they give the Union a naturally climate-sensitive external-geo-strategic projection. The EU's core environmental policies are not so unimpeachable that a lead role in the security dimensions of global warming can be taken for granted. The regulatory bent of much environmental policy even risks eclipsing the kind of strategic reference-frame needed for a climate security strategy. In addition, much policy-making effort remains focused on quite traditional parameters of energy security, in a way that is not obviously helpful to the profile accorded to climate security. Indeed, the priority the EU attaches to traditional aspects of energy security has recently begun to cut across climate security policies. Energy-related security is still deliberated as being overwhelmingly about guaranteeing oil and, increasingly, unconventional gas supplies rather than a need to pre-empt climate-induced instability. In both these areas – climate and energy policy, respectively – an assumption that extant EU competences naturally 'spill over' to provide a strong foundation for climate security is too easily adhered to. Climate change may have established itself firmly as a security issue on the EU policy agenda, but recent years have seen it fail to advance in tangible ways as much as other dimensions of European external relations.

Notes

1 S. Hammeling and D. Eckstein, *Germanwatch Climate Index 2013*, Bonn, Germanwatch.
2 Platts *EU Energy*, Issue 287, 13 July 2012.
3 European Commission, 'Key facts and figures on the external dimension of the EU energy policy', staff working paper, SEC (2011) 1022, pp. 7–8.
4 Figures from Platts *EU Energy*, Issue 252, 25 February 2011.
5 Platts *EU Energy*, Issue 264, 26 August 2011, p. 5.
6 www.gov.uk/government/policies/increasing-the-use-of-low-carbon-technologies/supporting-pages/carbon-capture-and-storage-ccs.
7 T. Bréchet, J. Eyckmans, F. Gerard, P. Marbaix, H. Tulkens and J. Van Yperselle, 'The impact of the unilateral EU commitment on the stability of international climate agreements', *Climate Policy*, 10, 2010, 148–66.
8 DG Energy, *Energy 2020*, Brussels, 2011, p. 5.
9 S. Fischer and O. Geden, 'Updating the EU's energy and climate policy', Berlin, Friedrich Ebert Stiftung working paper, 2013.
10 *Euractiv*, 22 May 2013.
11 European Commission, *Green Paper: A 2030 Framework for Climate and Energy Policies*, COM (2013) 169.
12 *Euractiv*, 27 March 2013, www.euractiv.com.
13 European Commission, *Green Paper*, p. 5 and p. 11.
14 N. Mabey, 'Europe has to follow its climate realpolitik', *ESharp*, 26 June 2011.
15 A. Bowen and J. Rydge, 'Climate change policy in the United Kingdom', Grantham Research Institute for Climate Change, policy paper, August 2011, p. 13 and p. 16.

16 D. Helm, 'What next for EU energy policy?', in K. Barysch, *Green, Safe, Cheap: Where Next for EU Energy Policy?*, London, Centre for European Reform, September 2011. Similar points are made by Gaventa and Mabey in the same volume.
17 Ibid, p. 5.
18 C. van Agt, 'The energy infrastructure challenge', in Barysch, *Green, Safe, Cheap*; Zachmann develops similar points in the same volume.
19 Platts *EU Energy*, Issue 246, 3 December 2010, p. 8.
20 http://ec.europa.eu/programmes/horizon2020/.
21 Ernst and Young, Renewable energy attractiveness index 2013, www.ey.com/Publication/vwLUAssets/Renewable_energy_country_attractiveness_indices_February_2013/$FILE/Renewable_energy_country_attractiveness_indices.pdf.
22 European Commission, *Green Paper*, p. 4.
23 Platts *EU Energy*, issue 276, 10 February 2012.
24 Platts *EU Energy*, issue 263, 29 July 2011, p.1.
25 R. Wurzel and J. Connelly (eds), *The European Union as a Leader in International Climate Change Politics*, London, Routledge, 2011.
26 K. Kulovesi and M. Cremona, 'The evolution of EU competences in the field of external relations and its impact on environmental governance policies', Rome, IAI, Transworld working paper 17, March 2013.
27 DG Energy, *Energy 2020*, p. 12.
28 European Parliament Foreign Affairs Committee, *EU External Policies for Future Energy Security*, EP, DG External Relations, 2012.
29 Article pending, 'Securitisation and development of a European energy policy', *Journal of International Relations and Development*. For overview, see C. Kuzemko, A. Belyi, A. Goldthau and M. Keating, *Dynamics of Energy Governance in Europe and Russia*, Basingstoke, Palgrave MacMillan, 2012.
30 C. Huhne, 'The geopolitics of climate change', speech at the Royal United Services Institute, 7 July 2011.
31 DG Energy, *Energy 2020*, p. 20.
32 www.counterbalance-eib.org.
33 C. Dupont and S. Oberthür, 'Insufficient climate policy integration into EU energy policy: The importance of the long-term perspective', *Journal of Contemporary European Research*, 8/2, 2012, 228–47, p. 238.
34 M. Luomi, 'Oil or climate politics? Avoiding a destabilising resource split in the Arab Middle East', Finnish Institute of International Affairs, briefing paper 58, April 2010.
35 S. Fischer and O. Gedan, 'The EU's energy roadmap 2050: Targets without governance', Berlin, SWP; *SWP Comments*, 8, March 2012.
36 Platts *EU Energy*, Issue 280, 6 April 2012.
37 M. O'Sullivan, 'The entanglement of energy, grand strategy, and international security', in A. Goldthau (ed), *The Handbook of Global Energy Policy*, Oxford, Wiley-Blackwell, 2013, p. 43.
38 European Parliament, *EU External Policies*.
39 DG Energy, *EU Energy 2020*, p. 20.
40 H. Winkler and J. Beaumont, 'Fair and effective multilateralism in the post-Copenhagen climate negotiations', *Climate Policy*, 10, 2010, 638–54, p. 641.

6

CLIMATE AND EUROPEAN DEFENCE STRATEGIES

Moving from the general set of European Union (EU) commitments into more specific policy domains, a first and vital consideration relates to how European armed forces have approached climate security. In several areas, recent years have witnessed a strikingly impressive uploading of climate concerns by European armed forces. Far from dismissing climate change as a mushy non-traditional threat unworthy of military attention, armed forces have on some measures assumed a lead role relative to other ministries. However, the inverse fear of militaries becoming overly dominant in climate security also appears unfounded – at least, so far. The sceptical perspective is that European militaries now regurgitate the standard discourse on climate security in purely disingenuous fashion. Critics warn that defence establishments have taken to frightening politicians that climate change is a security threat simply as a means to attract more funds for their own beleaguered bureaucracies.[1] There is some, but relatively limited, evidence that this is happening in Europe. Armed forces certainly seek to position themselves advantageously within climate security debates. But they have eschewed an unambiguous lead role in climate security relative to other ministries.

Militaries have so far engaged on a relatively narrow set of climate-related issues; they are still attempting to extend their grip on the broader geopolitics of global warming. Armed forces cannot be said to be leading climate security in the direction of undue militarisation. The most pertinent policy debate is the more subtle one of how far climate change may require a reconfiguring of defence to focus on 'protecting the homeland'. Militaries have not yet devised comprehensive rules and guidelines for their own operational doctrines that are capable of proactively tempering climate risks. While traces of the 'lifeboat' mentality have seeped into European planning, policy changes fall well short of any wholesale or qualitative shift in defence strategies.

Accepting climate security?

In a generic sense, European militaries have increasingly bought into the climate security agenda, despite it not constituting a traditional form of threat. The EU Military Staff has stepped up cooperation with the US Department of Defense. The North Atlantic Treaty Organization's (NATO's) *2010 Strategic Concept* mentions climate security as a key concern for the alliance's future remit. Its 'smart defence' concept has been defined to include climate-preparedness.[2] Yet, all militaries grapple with the question of how much priority to attach to climate relative to other aspects of security. They have incorporated the language and logic of climate security, but struggle to fully integrate this issue into their strategic planning.

A number of European armed forces have engaged seriously with the climate security agenda. Cooperation between European militaries on this issue is said to be better than on most other defence and security questions. The UK in particular has incorporated climate planning into its defence policy and introduced a more systematic coverage of climate change into its military staff colleges.[3] An influential agenda-framing role was played by rear admiral Neil Morisetti, as the UK's climate and energy security envoy (until 2013 when he was appointed as the government's overall climate envoy).

In 2010 Sweden's Ministry of Defence commissioned a strategy on how Sweden should adjust its national security and defence strategies to counter 'new and real threats', such as possible security tensions that may emerge from conflicts relating to the more aggressive development of natural resources by major powers in the Arctic region. The aim was to provide improved early warning and allow for better preplanning, for military crisis management operations and national defence readiness. The defence minister admitted:

> It is often said that there is a clear link between climate change and conflict as the natural resources base shrinks. However, there is relatively little research that deepens understanding of this, and we need more robust research to build systems that provide better early warning and preplanning. This applies to both civil and military crisis management operations.[4]

In its 2011 strategy document *Security, the Energy Model and Climate Change*, the Spanish Ministry of Defence committed itself to promoting further awareness of the impact of climate change on security and emphasising the need for multilateral cooperation between armed forces. The strategy emphasises the role that the military must play and the need for it to be fully conscious of how far climate issues will form a pivotal part of its future geo-strategic challenges; it claimed that the armed forces were already being remodelled with a new type of 'climate mission' in mind. Despite coming relatively late to the consensus view that climate change has repercussions on security, Spain's Socialist government became one of the most explicit of all EU states in committing its armed forces to multilateral operations in climate-induced scenarios. This commitment in part reflected Spain's relatively higher climate-security risk compared with other EU countries. Spain's vanguard role has been noted in its creating the aforementioned Military Emergencies Unit to respond to climate disasters. By 2012 this had undertaken 90 interventions, mainly within Spain itself.[5]

In many cases, however, the limits to armed forces' engagement have been especially notable. The German Defence Policy Guidelines 2011 contain very little on climate security.[6] German policy has not integrated the defence ministry as much as British strategy, in a conscious move to avoid a perception that there is a military lead on the new agenda. Germany has declined to appoint an equivalent to the UK's envoy that sits astride the foreign and defence ministries, preferring to keep the lead for climate security purely within the foreign ministry. Rather, the Auswärtiges Amt has consciously worked more with the development, environmental and education ministries.

While the Danish Defence agreement 2013–17 stipulates that the Danish armed forces are expected to contribute to the realisation of the government's goals within the climate field and commits to beefing up an Arctic preparedness force, the mandate is limited to the Danish armed forces reducing their own environmental footprint and energy consumption.[7] One semi-official Danish study concluded that Danish security thinking still lacks a 'comprehensive strategy' for rendering operational the well accepted agenda of climate security.[8] The 2009 Finnish Security and Defense Policy contains two brief pages on climate security that forward little in the way of tangible policy change. The Netherlands Defence doctrine does not contain any particular reference to climate challenges.[9] The National Security Strategy of the Republic of Poland of 2009 is bereft of operational specifics on the climate issue.[10] The 2011 Security Strategy of the Czech Republic identifies climate security as a priority but in a generic sense and not linked in an identifiable way to changes in defence strategy.[11] Other member states, such as Italy, Belgium, Portugal, Greece and Estonia, have declined to flag up the issue publicly in their strategy documents.

Notwithstanding their ministry's new deliberations, Spanish officers themselves acknowledge that a full integration of climate change factors into strategic planning remains a long way off. The ministry's 2011 document acknowledged 'grave deficiencies' in the levels of equipment needs to project external deployments for climate-related conflict. Senior generals are reported as admitting that the Spanish military has made relatively little progress in delineating an overarching strategy that specifies the Ministry of Defence's role in relation to climate change. After landmark initiatives committing the armed forces to be an environmental protection force, the Spanish defence ministry has exhibited little follow-through on such initiatives.[12]

A French military spokesman admits: 'we are only at the beginning of looking at this.' France has introduced a few new climate-and-security projects but, insiders acknowledge, has not gone far in really according climate security top priority at a political level; it is seen as an issue only amongst a confined circle of the 'initiated'. The struggle for resources for this agenda was a difficult one in the best of times, but has become even more challenging in the midst of economic crisis. The UK's envoy recognised that counter-terrorism and other traditional security threats command the government's ear far more readily by virtue of being able to point to a tangible threat and specifiable risk if action were not taken – things much more difficult to pinpoint in the case of climate security. Already-running military operations are given even more unquestioned priority. An EU military staff officer reports that the prevailing view is that climate security merely strengthens the case for the Union's existing 'comprehensive approach' to security.

In sum, military thinking has evolved but within fairly narrow parameters. European militaries are certainly not head-in-the-sand climate-deniers, but neither have they led a genuinely overarching geo-strategic rethink. They tend to think in terms of what kind of new kit they will need, but not really of how underlying causes must be factored into their preventive diplomacy and readiness.[13] European defence ministries are now fully engaged on the question of how climate change is likely to impose new requirements and conditioning factors on their own operations and immediate modus operandi. Their senior officials admit that they are still in the process of taking the next step to envisioning how climate change could act as a powerful modifier of geopolitics and thus impinge upon the broader contours of defence policy. Militaries have not approached climate security in a purely disingenuous fashion – simply to dress up their own, familiar resourcing concerns in environmental discourse. But it does mean that so far they have judged that their own role is pertinent to a relatively confined part of the climate security jigsaw.

There is general agreement that European defence establishments lag behind the US military's engagement with climate change issues. Significantly, while the US has, of course, been a laggard on climate change in general, in its securitised realm Washington defence planners remain steps ahead of their European counterparts. Under the 2007 Global Climate Change Security Oversight Act, the US has initiated a far more systematic programme of research on the impacts of global climate change on military requirements, operations, doctrine, organisation, training, material, logistics, personnel and facilities, and the actions needed to address such impacts. In June 2010, the director of the US Navy's Task Force on Climate Change stated:

> We want to basically pace the threat. We don't want to get into a tail chase over climate change, but at the same time, we do not want to spend ahead of need, spending for things that may not be required for years or decades later.[14]

The 2010 Quadrennial Defence Review talks of climate change as an 'accelerant' of instability, and in general accords the issue higher profile than most European defence documents.[15] It is often now referred to by US planners as a 'ring road' issue, through which all other security issues must pass.

Towards 'climate interventions'?

Some observers detect significant movement: even if a purely climate-related common security and defence policy (CSDP) mission may be unlikely still, EU planners have begun to assess climate factors as part of conflict management scenario-building, as well as tighter coordination with the EU's Civil Protection Mechanism.[16] However, while European militaries have begun to take climate security seriously, there is little evidence of preparedness for armed interventions being carried out systematically as a central part of this agenda. Sceptics have long feared an over-militarisation of climate issues; in practice, militaries and the wider EU security establishment remain extremely circumspect. At least for the moment, the prospect of EU military missions being deployed regularly as a response to climate change events appears a distant one.

The European Parliament's October 2012 report was largely negative in its assessment of EU policies implemented to date. Climate security has been 'overshadowed' by the eurozone crisis and climate-sensitive assessments have not yet been incorporated into CSDP planning. No list of climate-stressed states has been produced in which the EU should prepare to act; no matching of generic threats to actual EU policies has been forthcoming (with the report stressing that the United Nations's (UN's) Environment and Security initiative has similarly failed in this). Climate impact is woefully short of being mainstreamed into EU defence coordination in the way that gender or human rights have been. The External Action Service (EAS) has failed to take advantage of the new Lisbon Treaty provision that enables it to launch a start-up fund as a means to enhance available capacities to respond to climate conflicts and disasters; indeed, defence cutbacks seriously compromise the

Union's ability to intervene usefully now. The EP report found that the large number of EU documents that have given operational content to conflict management – including the EU Concept for Military Planning at the Political and Strategic level, the EU Concept for Military Command and Control, the EU Concept for Force Generation and the EU Military Rapid Response Concept – are bereft of climate-related considerations. Additionally, EU delegations around the world lack expertise on climate security.[17]

European armies are still geared towards territorial defence against an attack, not mobilising for conflict operations and complex crisis-disasters in places like Africa or the Middle East. The EU defence spend is heavy on personnel and low on high-end capabilities.[18] European Union member states have nearly 2 million personnel under arms but only 100,000 of these are deployable to conflict zones. A full half of EU deployable capacity comes from the UK, reflecting the limited investments made by other states. The EU's Battlegroups became operational in 2007; by 2013 they had still not been used. They will be useful only for basic crisis-moment protection – controlling airports, evacuating European citizens – and do not offer the prospect of being integrated into longer-term rebuilding and development efforts.[19] Experts lament that rather than being used as a de facto strategic reserve to help sustain long-term peace-building strategies, the Battlegroups have been conceived as vehicles only for the one-off crisis response.[20] The EU's understanding of a 'rapid' reaction force is, at the very least, elastic: the 2008–09 Chad mission was delayed eight months partly due to a shortage of transport and logistical equipment.

The twenty-eight CSDP missions carried out to date have not been ambitious and have been mostly tangential to core EU interests. Most of the missions have constituted little more than symbolic gestures. Only six CSDP missions have had a military component: those in Macedonia, DRC (in 2003 and 2006), Bosnia and Chad, in addition to the 2009 anti-piracy mission off the coast of Somalia. Only seven have involved over 200 personnel. The EU has limited itself to supporting 'multilateral subsidiarity', through support for the interventions of other organisations, particularly the African Union in the cases of Somalia and Sudan.[21] The focus has been on very short crisis interventions. It is doubtful that CSDP can be seen to have meaningfully enhanced the EU's implementation of 'Responsibility to Protect' commitments.[22]

A number of CSDP missions have been deployed to climate-stressed areas where environmental factors are seen as contributing to instability. These include the maritime mission Atalanta off the Horn of Africa, and security training initiatives in the Sahel (especially through missions in Niger and Mali in 2012 and 2013, respectively). Many policy-makers see such deployments as a harbinger of future defence requirements. The EU has invested heavily in a Global Monitoring for Environment and Security (GMES) system that is now being rolled out, with a range of satellites and other capacities. This has given some back-up to interventions in Pakistan and Haiti. However, while such missions and support reflect an evolution in military thinking, they have been extremely low key and cannot be said to represent major deployments triggered primarily by climate factors. No EU military deployment has been countenanced to safeguard supplies coming into Europe, as one might expect in a 'lifeboat' security scenario.

The UK's climate security envoy stressed that climate change would begin to shape UK military deployments and missions. However, while the British military has begun to think in terms of climate security, tangible and operational changes remain limited. The UK's Strategy and Defence Review contains nothing on reconfiguring the military for a climate-threatened international system. Those close to the centre of planning decisions reveal that the priority attached to resourcing existing missions is delaying the necessary military reconfiguration appropriate to climate change impacts. It has proven impossible for ministers, under intense public pressure for under-equipping troops in theatre, especially in Afghanistan, to use scarce resources for nebulous long-term threats rather than plugging the gaps in basic operation equipment. As forces are removed from Afghanistan, the conviction is that more scope will exist to focus on longer-term climate threats.

In 2007, the British Ministry of Defence (MoD) signed a three-year, £12 million deal with the Met Office Hadley Centre designed to identify where conflict and security threats might emerge as a result of climate change.[23] However, in 2009 it retracted over a third of that budget, justifying the cut as a necessary diversion of resources to 'prioritising success in current operations, such as Afghanistan'.[24] The programme was not renewed after 2010. Officials complain that in a time of shrinking overall defence spending, the easiest cuts to make are those that relate to preparations for longer-term risks. While the MoD may have come to accept the need to take climate change seriously as a long-term risk factor, its officials tend to think in a relatively confined way about how environmental pressure will condition military missions; they have done little to suggest that they conceive climate change as a factor that is likely to condition the broader politics of international security and the UK's strategic positioning. In private, diplomats admit that they are only able to paint the very broad impacts of climate change at a regional level and do not have the detail to understand country level effects in a way that reshapes concrete conflict interventions and peace work on the ground.

As commented, Spain's Unidad Militar de Emergencias (UME) was established specifically to deal with natural disasters and environmental catastrophes, including those triggered by climatic factors. The unit consists of 4,000 personnel, and answers to the minister of defence. The UME was activated in Haiti, during the Lorca earthquake relief operation, and in response to fires in Galicia and the Baleares. The UME has been studied as a model by officials from the US, China and Latin American countries.[25] However, while it is true that the UME deployed effectively to Haiti and that Spain led the NATO mission that was sent to deal with the impact of floods in Pakistan in 2009, the Spanish military's consideration of climate security has so far been limited to two issues: responses to such natural disasters and the need for defence forces to improve their own energy efficiency.[26] Spanish defence officers say they understand the climate issue as being mainly about making logistics lighter for field missions.

Sweden has begun to give thought to the role of its armed forces in multilateral operations, and in particular of how its military will require greater training in and knowledge of climate factors. As one official reports: 'We are already looking into the effects of carrying out civilian missions and military operations in climate-stressed environments.' As one example of progress, the Swedish armed forces have incorporated teams that can assess the environmental situation on the ground and ensure that the military itself does not cause harm as it moves into fragile, at-risk areas.

UN representatives and development agencies still fret over an encroaching military role, accusing the armed forces of seeing climate conflict as a problem to be contained through armed intervention rather than an issue about social empowerment to attain long-term solutions. European military representatives deny this strenuously and insist they do not want the lead role on climate security. They claim that their role is a relatively minor one and that – contrary to critics' accusations – it does not conveniently help them mobilise resources as it is not a high-profile enough issue for governments. In fact, militaries feel repelled by other actors' perspectives on climate security. Militaries complain that these other actors have increasingly couched climate security in a language unfamiliar to the armed forces with little emphasis on the kind of impacts that have the tangible, discrete problem-solving quality that militaries like.

Defending the homeland?

Some observers judge that the most practical change in the field of climate security to date has been the pressure on governments to turn armed forces inwards to deal with domestic floods and storms; in consequence, the danger is emerging of climate change actually weakening the focus on foreign policy engagements.[27] Several analysts have suggested that climate change's most notable impact on military configurations will be the onus it places on defending home territories against extreme weather. One expert on military strategy notes the trends amongst defence planners: a 'renewed interest in a nation's civil defence capacity' and in boosting national guards; and a re-reorganisation of forces to keep as many troops as possible stationed out of the regions likely to require climate-related intervention.[28]

While much of the new EU security rhetoric is about addressing climate-ravaged failed states, in reality pressure will be strong for resources to be diverted into protecting Europe itself from natural disasters and climate-related stresses. Some detect trends in military spending towards transforming military forces into something akin to home guards with the task of defending domestically against climate ravages. This is part and parcel of the kind of 'lifeboat mentality' that many experts have predicted and indeed advocated. It also chimes with the cuts in defence expenditure made by nearly all European states in the wake of the post-2008 financial crisis. The decrease in defence budgets is leading many in European armed forces to argue that they should focus more narrowly on core defensive functions to avoid dangerous overstretch.

The EU adopted an Adaptation Strategy in April 2013 that was focused on such internal initiatives. This predicted that differential effects across the Union would undermine European unity; that the cost of not acting would amount to €250 billion per year by 2050; and that climate effects could cause 90,000 deaths in the EU by the same year. The strategy recognises that internal adaptation measures remain at 'an early stage' in Europe. Fifteen member states have national adaptation strategies that are just beginning to engage in meaningful projects. European Commission efforts at coordinating these national efforts are even more embryonic. The new strategy commits the EU to better exchange of member states' best practice; peer pressure to ensure all member states implement national strategies; awareness-raising across Europe; more research; capacity-building projects; and EU-wide vulnerability assessments, culminating in 'comprehensive threat and risk assessment reports' to be produced jointly by the Commission and High Representative in 2015. The strategy promises more funding for adaptation under the EU's new 2014–20 budget, albeit without specifying amounts.[29]

In 2011 4.4 per cent of France's defence budget was spent on the category 'Environment and future defence policy'. The Dutch have invested several hundred million euros in strengthening coastal defences. In 2012, 2.2 per cent of Denmark's defence budget was allocated to Home Guard Command and another 2 per cent to the Emergency Management Agency. While Sweden's defence policy does not include any particular reference to defence strategies adopted for climate change challenges, a clear focus has been on preparing to deal with floods.[30] The most notable feature of military operations may not be a renewed commitment to the flagging 'fragile state' agenda but rather a more direct defence of key energy supply lines within Europe. It is often pointed out that the infrastructures necessary to bring renewable energy imports into Europe are likely to be just as, if not more, vulnerable than oil and gas pipelines. Similarly, the UK has bases on low-lying coral atolls in the Indian Ocean; the loss of Diego Garcia to rising sea levels would weaken British influence with the US, to whom the base there is leased.

The UK government has advanced a plethora of defensive adaptation measures, introducing a range of policy initiatives, including Climate Change Risk assessments, a National Adaptation Plan, Climate Resilience Toolkits for businesses, and several billion pounds spent since 2010 on flood and coastal erosion adaptation. Local authorities in London have prepared a number of reports on how the capital's transport and other infrastructure systems can be protected from climate change factors such as increased flooding. Fundamental reforms to the UK military in recent years have been ostensibly designed to focus more on preventive action, including on climate preparedness. This will involve pulling troops back from overseas engagement for a more defensive 'contingency' defence doctrine, while reducing the military to its smallest size for more than a century.[31] The UK's outgoing climate security envoy argued that threats to the global trading system have a read-over to military configurations: the notion of 'maritime resilience' has taken root, reflecting the importance of missions aimed at keeping commercial sea routes open and free of conflict. UK officials talk of the military being mobilised to protect sea lanes, communications and trade routes. A Royal Navy ship on tour in the West Indies now gets disaster relief training. UK officials have made trips to the US to learn about preparedness for homeland defences.

Yet, on homeland defence the EU also lags well behind the evolution in US military planning. In the US the focus has been on shoring up both military and civilian capacities to deal with disasters in the United States itself. This was a result of the tragic under-preparedness revealed in the wake of Hurricane Katrina. Even here, where the 'homeland defence' perspective has been so clearly pre-eminent, critics argue that the progress has been insufficient. While there have been some similar steps adopted in Europe, this kind of policy orientation has been much less evident than in the US.[32] Member state diplomats acknowledge that homeland defence is still not advancing significantly in Europe. Only a quarter of US troops are trained for disaster relief; a big effort has been made to increase this proportion. Much less investment has been forthcoming in Europe. This type of reconfiguration has been more notable in the US than in Europe. US strategic planning has been much more tightly oriented towards guarding supply routes and shipping lanes.

Notwithstanding the steps adopted to enhance protective measures, UK policy-makers are adamant that reconfiguring defence capabilities on the basis of a 'lifeboat Britain' scenario is not feasible as a long-term solution. Debate has taken shape on such an approach – certainly in the UK, where by dint of geography the lifeboat scenario at least appears as an option. Some effort has certainly gone into thinking what kind of military resources would be needed to turn back growing hordes of environmental refugees from British shores. As yet, however, there is no discernible preparation for such a lifeboat-based approach to policy. A lifeboat strategy would require some fundamental choices, such as withdrawing from the EU or the World Trade Organization (WTO) and many other international bodies and conventions; politicians insist they are not willing to entertain such drastic steps, at least as yet.

Greening the military

The European defence establishment has begun to demonstrate environmental awareness by shifting away from fossil fuels to more sustainable sources and reducing energy consumption. The arguments used are not purely about climate change and environmental responsibility, but rather about national security. This relates especially to tackling fuel dependence which is seen as increasingly expensive and strategically undesirable. Many interesting 'greening war' initiatives have been supported. In June 2012 the European Defence Agency launched a Military Green programme of procurement to coordinate the plethora of member state plans in this area.

Britain's ministry of defence is responsible for 1 per cent of UK carbon emissions. It has in recent years stressed a commitment to reducing these emissions and 'making war green'. In a 2011 strategy document covering this question, the ministry insists that 'There are moral, legal and financial reasons why the MoD has to strive to make its equipment environmentally robust throughout the lifecycle.'[33] Under UK legislation the MoD has a legal duty to ensure that its activities do not cause environmental damage. The ministry perceives increased investment in low carbon technologies as necessary to reduce the amount of fuel being taken to the front line, which has become an ever-more costly, risky and logistically resource-intensive supply-line impediment.[34] Neil Morisetti has argued:

> Identifying ways to reduce our emissions also helps improve our operational efficiency in theatre … Transporting fuel puts additional lives in danger, and greatly increases the financial cost of operation. Likewise, reducing the hull friction of our warships has delivered improvements in fuel efficiency.[35]

The French defence ministry has invested nearly €10 million a year into research designed to make its military equipment greener and more energy-efficient.[36] The ministry allocated €60 million in 2010 under its Environmental Action Plan for Defence. Measures included the aim to have 80 warships awarded a 'green passport' by 2012 and 31 armament programmes comply with a series of eco-standards. Spain's Ministry of Defence has introduced a new Energy Efficiency and Savings Plan which commits to reducing the military's energy consumption by 20 per cent by 2016.[37] In 2011, funding of €27 million was forthcoming under this plan for a range of military-related environmental projects. The ministry also supported a new masters course on the environment, specifically aimed at training personnel across the three branches of the armed forces and the civil guard on climate change debates.[38]

Some recent academic accounts insist that, in practice, militaries have failed to meet their own 'greening' commitments and remain amongst the worst polluters worldwide.[39] Moreover, notwithstanding all the formal commitments, European officers and officials recognise that the US military has, once again, been ahead in setting the terms of this area of debate. The US military is the biggest user of energy in the US, accounting for 1 per cent of the country's energy consumption, and so has had a significant incentive to reduce its own consumption. In 2010, the US Navy launched the F-18 Green Hornet fighter jet flying on a biofuel blend. The US is set to launch its Great Green Fleet initiative by 2016, comprising a fighting force of ships, submarines and planes powered entirely by biofuels. Testing and evaluation of alternative fuels have been extensive, for example in the 49-foot fast and agile RCB-X boat. The military is committed to procuring 25 per cent of its electricity from renewables by 2025. US Navy rear admiral Philip Cullom argues that,

> Going green is about combat capability and assuring the Navy's mobility. It is not just about natural security; it also strengthens national security. By having reliable and abundant alternate sources of energy, we will no longer be held hostage by any source of energy, such as petroleum … First and foremost, energy conservation extends tactical range of our forces while also preserving precious resources. Our goal, as a Navy, is to be an 'early adopter' of new technologies that enhance national security in an environmentally sustainable way.[40]

Such advances are despite a warning from the RAND corporation that in practice, 'The use of alternative fuels offers the armed services no direct military benefit.'[41]

Shifting security alliances

While militaries have got used in recent years to dealing with climate-induced disasters, it is not clear that they have extracted any lessons from these that are qualitatively different from other types of disaster. Militaries admit they are not using climate-related interventions as a base from which to build partnerships with other states and armies to advance cooperation on the broader geopolitics of climate security. For example, European defence officers still need engagement with militaries in the many developing states where they – the military – are the ones responsible for exploitation of forests and other habitats. As the need for interventions increases, European governments have done little to attain positive signals towards burden-sharing from other states. Relatively limited military-to-military links on climate issues have been established with other powers.

European militaries still tend to elide the 'security implications of climate security' with 'threats' rather than approaching the issue in a broader sense of power balances needing to be reconsidered and reshaped. Militaries have not engaged strongly with or pushed for mitigation, but have rather limited themselves to reflecting on how to 'pick up the pieces' when disasters and conflicts strike. They admit to being at a relatively early stage in trying to define what the security aims actually are in this area, beyond effective responses to climate-induced disasters. This reveals a key point: there has been a strong focus on the potential for *intra*-state conflict (as we will see in the following chapter), but very much less on possible *inter*-state tensions within a climate-stressed and fluid global order.

In general, European defence establishments insist strongly that they support deeper multilateralism as a core organising principle for responses to the climate security agenda. A constant theme that emerges in conversation with European diplomats and military officials is the concern to downplay rivalry-based international scenarios. All European documents and officials stress that climate change should not be painted as a 'north versus south' issue but a common global problem requiring multilateral partnerships. It is difficult to discern any significant impact on European defence preparedness of sceptical analysts' predictions of climate-driven zero-sum conflict between major powers. European militaries themselves recognise the lack of progress on factoring into the geo-strategy calculus incipient changes in the broader international dynamics of conflict. It is notable that military staff within the EAS have not been prominent players in the design of EU policy. Indeed, there has been negligible discussion between CSDP units and other parts of the EAS about climate-related geo-strategy. Only in the wake of the new 2011 climate diplomacy strategy did military staff begin to show an interest in engaging with other parts of the EAS in this question. In 2013, this has developed much further, with a major new programme involving defence staff in climate-related planning.

The UK government Foresight report of late 2011 registered this point unequivocally: the long-term impact on UK interests will be most serious not at the level of military intervention but the potential dysfunctionality of global institutions. If the kind of multilateral institutions that have served Western interests well were to atrophy in the face of climate security challenges, the advent of a more zero-sum security panorama could leave Western powers severely weakened. While the UK government has begun to build climate risk factors into its scenario building for humanitarian response planning, the broader implications for geopolitics are acknowledged to have fallen outside the purview of military and other strategic planners.[42] British diplomats argue that too much has been expected of the UNFCCC alone, and that the latter needs to be complemented by other, more geopolitical fora.

European strategic cultures show limited signs of addressing the broader international factors that play into risk calculations. For example, one danger is that climate change will encourage Asian states to build up militaries to have available the equipment to defend against climate insecurity, and that this will have deleterious spill-over effects in terms of a general militarisation of regional dynamics. The ASEAN Regional Forum has already held mock military exercises to deal with region-wide environmental disasters; the US but not the EU was invited to participate. The EU and ASEAN tried but could not succeed in agreeing a common declaration on climate change. The 2012 EU–ASEAN summit achieved little in this area: climate security was not prominent in the Enhanced Action Plan agreed between the two regions. European officials stress that they have made little headway in getting Japan and Australia to partner on climate security initiatives, although Korea has been keener on military-to-military cooperation.

As outlined in chapter four, particular effort has been made to develop a strategic alliance with Beijing; indeed, this has relegated the importance of other areas of policy in relation to which the EU and China have for many years not seen eye to eye. It is widely recognised that in climate policy, all other challenges pale alongside the need to cooperate with China on low carbon. An EU–China Partnership on Climate Change embraces a range of activities related to clean energy technologies. A more specific and targeted China–EU Action Plan on Energy Efficiency and Renewable Energies promotes industrial cooperation relevant to protecting the global environment. A biennial EU–China energy conference brings together high-level representatives from European and Chinese industries and governments.[43] A big EU–China CCS initiative is seen as such a promising model that the EU has been keen to extend a similar initiative to India. However, diplomats acknowledge that the hard security dimensions of climate change have still not been incorporated into the EU–China relationship, beyond generic conversations about potential threats ahead. Catherine Ashton has commenced a new regular strategic dialogue with Chinese officials, and has pushed hard to incorporate climate security issues into this. In general, however, China robustly rejects any kind of security multilateralism in its region; EU member states for their part admit that trade and investment with China have been by far the overwhelming concern during the eurozone crisis.

Where concrete hard security dangers have begun to manifest themselves, European defence and strategic planners have struggled to influence overall EU policy in any tangible fashion. By way of illustration, one such apparently imminent flashpoint derives from Egypt's new water plan, that risks causing massive instability upstream in the Nile Basin across several neighbouring states: the upper Nile states now insist that they want to renegotiate terms and take more water from the Nile. The Nile Basin accord is breaking down as Egypt refuses to talk to Ethiopia about reallocating water allowances (although there were attempts at restarting dialogue at the end of 2013). As one security official acknowledges: this is the kind of conflagration-in-waiting that the EU needs to help avert, but whose delicate political context renders any outside security engagement extremely delicate. The EU has upgraded its relations and cooperation with Egypt in the wake of Hosni Mubarak's ousting; but one searches in vain for any engagement on this festering security risk. Years of controversial cooperation with the Egyptian military appears to have given European governments little leverage on these kinds of climate security challenges. The EU continues to contribute to security in this region by deploying ships off the shore of Somalia to deal with small groups of pirates; an important challenge but marginal to the game-changing resource stresses soon to be felt. The most common view expressed in private by EU security officials is that a crisis-enfeebled Europe cannot realistically hope to engage with notable influence in the large number of serious cross-border climate hotspots. The most that EU defence capability can aspire to, many of them insist, is a more systematically climate-sensitive contingency planning.

The Arctic is one arena where alliances have shifted and the EU has tried to respond. Prior to the 2012 strategy document, detailed in chapter four, Catherine Ashton travelled to the Arctic, eager for the Union to receive permanent observer status on the Arctic Council, 'because some of our member states are so connected to it, [and] … because all of the European Union has something to offer, when you think about some of the environmental issues'. The European Parliament passed a resolution on Arctic governance in 2008, pushing the EU to take the lead in drawing up a multilateral treaty to govern the Arctic; this suffered a hostile response from the Arctic powers that pointed to existing arrangements working well and not needing to 'import' EU rules. The EU's ambitions have been lowered to providing small parcels of technical cooperation and ensuring that the EU at least has a seat at the table.

While it is true that the Arctic states have all begun to reconfigure their militaries to safeguard their interests in the region, so far they are cooperating in a relatively positive fashion and the Arctic Council is working. The EU has been trying to stake a role in mapping the incipient Arctic disputes, but so far the participating states insist that their own mapping is proceeding smoothly. It may be that the most notable development is not so much geopolitical tension in the Arctic but rather the Arctic Council taking on the role as an increasingly powerful club; Scandinavian states may begin to feel much more epistemic identity here than with southern EU states.

Danish diplomats insist that the country maintains a policy based on positive-sum cooperation rather than military defensiveness, to avoid low-level differences becoming a major security problem. The Arctic Council has awoken after years of slumber, pushed by Denmark and others into a more proactive role. Denmark has resisted any pressure to have NATO involved, with the argument that this 'would create a security problem where none currently exists'. The main Danish base in Greenland has indeed been built up and Denmark is sending and equipping more military frigates for the Arctic. But the government insists this is to cover coast-guard search and rescue functions and also to prepare for new economic opportunities in the region – with Danish registered ships carrying 10 per cent of global shipping, the opening of new trade routes is seen as a potential advantage.

If the Northwest Passage opens up, a crucial question will be who is responsible for policing the new route. Rejecting a defence-led strategy, the EU has proposed a shared governance approach to the Arctic, rather than support Canada's claim to police the new route – even if, as mentioned, some now see it as backtracking somewhat from this position. Arguably, the EU has not approached its relations with Iceland in a 'hard security' sense either, immersed in detailed financial differences. Critics accuse the EU of simply trying to justify a role for itself in the Arctic, thinking its own rules can serve as a model for its management, when Russia, China and other powers are aiming to gain control in a much more directly political sense.[45]

Conclusion

The military has the potential to act either as a powerful catalyst for a serious climate security strategy or to complicate progress. Earth Security Initiative director Alejandro Litovsky writes that, 'When the military talks climate change, everyone listens'; he argues that this new interest could either lead to a win–win gain for both security and environmental communities, or if badly handled to an unnecessary escalation of instability on the back of climate change policies.[46] This chapter reports that the policy evidence suggests the jury is still out on this question. Policy-makers agree that there has been progress in getting the military to 'think green' but that similar advances have not been forthcoming in how climate change relates to broadened definitions of security.

The military focus is most strongly focused on the armed forces' own energy consumption and how to run operations in climate-stressed environments. Efforts have been more halting to go beyond 'green procurement'. European militaries are more geared towards dealing with extreme weather events than they are seized of the broader climate-engendered changes to geopolitics. Moreover, on their own admission militaries still think in terms of the 'national state interests' to be defended against climate change, not the broader impact on human security – and not the need proactively to address the root causes of individuals' rights and livelihoods. The argument that governments and militaries have used climate security as a means of boosting realpolitik hard-security policies is of some, but relatively limited, use in explaining the evolution of European policies in this area. If there was initially a military drive behind the climate security agenda, now the EU arguably risks going to the other extreme of approaching the challenge as one of micro-level adaptation and not of a more political-level reshaping of international power balances. Ironically, militaries are berated for jumping on the climate bandwagon but also for restricting themselves to one modest part of the climate security equation. It is reassuring that climate security has not (yet) been overly militarised and that militaries have kept their engagement on the issue within reasonable parameters. It is less admirable that broader reflection and planning on defence reconfiguration remains so patchy across Europe.

Notes

1 P. Khanna, *How to Run the World: Charting a Course to the Next Renaissance*, New York, Random House, 2011, p. 10.
2 NATO, *The 2010 Strategic Concept Active Engagement, Modern Defence*, Brussels, NATO, 2010.
3 Her Majesty's Government, Ministry of Defence, *Climate Change Strategy, 2012*, www.gov.uk/government/publications/mods-climate-change-strategy.
4 Gerard O'Dwyer, 'Sweden commissions climate-based security survey', *Arctic Security*, October 2010.
5 Instituto Española de Estudios Estratégicos, *Seguridad, Modelo Energetico y Cambio Climatico*, Madrid, Ministerio de Defensa, 2011, p. 184, p. 213, p. 217 and p. 227.
6 www.bmvg.de/portal/a/bmvg/!ut/p/c4/LYsxEoAgDATf4gdIb-cv1MYBzcQbMDgQ8ftSONtssUsrddQ3iDdk9YlmWnaM4XXhauIq9pPLybB65wRDdF6FQzZ2R47PxdqtcTHGAXlU_q72byv9tgQFK91xGj6tRgx1/.
7 www.fmn.dk/eng/allabout/Documents/TheDanishDefenceAgrement2013–17english-version.pdf.
8 S. Nordstrom, *Climate Security: From Agenda-setting to Policy*, Copenhagen, Royal Danish Defence College, 2010.
9 www.defensie.nl/english/organisation/ministry_of_defence/defence_doctrine/.
10 www.archiwalny.mon.gov.pl/pliki/File/English/strategia_obronnosci_eng.pdf.
11 www.army.cz/images/id_8001_9000/8503/Czech_Security_Strategy_2011.pdf.
12 Ministerio de Defensa, 'Fuerzas armadas y medio ambiente', *Revista Española de Defensa*, 226, 2007.
13 H. Brock, *Climate Change and Drivers of Insecurity in the Global South*, Oxford Research Group, 2012.
14 www.defense.gov/news/newsarticle.aspx?id=59713.
15 Department of Defense, *Quadriennial Defence Review 2010*, www.defense.gov/qdr/images/QDR_as_of_12Feb10_1000.pdf, p. 85.
16 F. Ducrotté, 'The impact of climate change on international security: Prospects for an environmental dimension in CSDP missions', ISIS Europe, *European Security Review*, November 2012, p. 6.
17 European Parliament Foreign Affairs Committee, *The Role of the CSDP in Case of Climate-driven Crises and Natural Disasters*, EP report, October 2012, p. 4 and pp. 7–8.

18 D. Keohane and T. Valasek, *Willing and Able? EU Defence in 2020*, London, Centre for European Reform, p. 18, p. 20 and p. 37.
19 C. Mölling, 'EU Battle Groups 2007: Where next?', *European Security Review*, 31, December 2006, 7.
20 *European Security Review*, ISS Europe, 39, July 2008, 3.
21 M. Gibert, 'Monitoring a region in crisis: The European Union in West Africa', Chaillot paper 98, January 2007, p. 30.
22 For an overview of ESDP developments and issue of capabilities, see Keohane and Valasek, *Willing and Able?*; N. Witney, 'Re-energising Europe's security and defence policy,' ECFR policy paper, July 2008.
23 http://webarchive.nationalarchives.gov.uk/+/www.mod.uk:80/DefenceInternet/DefenceNews/DefencePolicyAndBusiness/MetOfficeClimateChangeStudyCouldHelpIdentifyFutureSecurityThreats.htm.
24 www.nature.com/news/2009/090625/full/news.2009.602.html.
25 www.defensa.gob.es/ume/noticias/2011/07/Noticias/2011_07_19_camp_verano.html.
26 I. García Sánchez, 'El cambio climático: implicaciones para la seguridad y la defensa', in Ministerio de Defensa, *Seguridad*, pp. 208–12.
27 A. Froggart and M. Levi, 'Climate and energy security policies and measures: Synergies and conflicts', *International Affairs*, 85/6, 2009, 1129–41.
28 Commodore S. Jermy, *Strategy for Action*, London, Knightstone Publishing, 2011, p. 148.
29 European Commission, *An EU Strategy on Adaptation to Climate Change*, COM (2013) 0216 final.
30 FOI (Swedish Defense Research Agency), *On Connecting Climate Change with Security and Armed Conflict*, Stockholm, Swedish Defence Research Agency, September 2010.
31 *Economist*, 22 October 2011, p. 43.
32 C. Paskal, *Global Warring*, Basingstoke, Palgrave Macmillan, 2010, p. 45.
33 Ministry of Defence, *An Introduction to Environmental Management in the MOD Acquisition Process*, London, The Stationery Office, 2011.
34 www.mod.uk/NR/rdonlyres/378271EE-0F39-4DF2-8FBB-E56E42733AD6/0/DefenceinaChangingClimateFINAL.pdf.
35 blog.decc.gov.uk/2011/02/11/a-military-response-to-climate-change/.
36 www.diplomatie.gouv.fr/fr/IMG/pdf/Brochure_climat.
37 www.defensa.gob.es/gabinete/notasPrensa/2011/02/DGC_110215_ministra_presentacion_cuadernosIEEE_CESEDEN.html.
38 http://webcache.googleusercontent.com/search?q=cache; www.defensa.gob.es/Galerias/areasTematicas/medioAmbiente/fichero/DGN_IniciativasMedioambientales2010-11.pdf.
39 M. Baranagan, *Global Warming, Militarism and Non-violence*, Basingstoke, Palgrave, 2013.
40 www.navy.mil/search/display.asp?story_id=56757.
41 www.rand.org/pubs/monographs/MG969.html.
42 Foresight, *International Dimensions of Climate Change*, London, The UK Government Office for Science, 2011, p. 45.
43 For information on this EU–China partnership, see House of Lords European Union Committee, *Stars and Dragons, The European Union and China*, London, The Stationery Office, 2010.
45 Paskal, *Global Warring*, pp. 124–25.
46 A. Litovsky, 'Militarising climate change', *Earth Security Initiative Comment*, 6 July 2011, www.earthsecurity.org.

7

CLIMATE AND EU CONFLICT PREVENTION POLICIES

The European Union (EU) has a rich experience and strong presence in the area of conflict prevention. As indicated in the previous chapter, internal civil conflict has been increasingly linked to climate factors to a much greater extent than have deliberations over broader conflict *between* states. The pertinent question is whether the extensive range of European conflict policies has changed in response to the perception that climate change is likely to intensify civil conflict. Nick Mabey has made the astute point that addressing the climate–conflict link requires governments to do more than simply add a little extra effort to their existing, generic conflict prevention policies. Rather, they need to tailor these policies much more specifically and tightly to the nature of the instability likely to result from climate change.[1] While the EU and member states have incorporated some climate-related dimensions into their conflict prevention instruments, their moves in this direction have lacked conviction. European policy-makers most commonly opine that climate change is indeed likely to augment the risk of civil conflict in resource-stressed societies, but also acknowledge that this prognosis has not prompted any significant upgrade in EU conflict prevention efforts. Nor has it led to qualitatively different approaches to conflict that discernibly build from climate-related risk indicators. The EU's 'access to energy' initiatives demonstrate some promise of introducing such tailored specificity, but concerns persist that these may not be entirely consistent with broader climate change objectives.

Climate and EU conflict prevention

Recent years have witnessed a general tailing-off in European political commitment to conflict prevention. The prospective threats of 'climate conflict' have not sufficed to reverse or temper governments' diminished investment in external security capabilities. Notwithstanding the ubiquitous predictions of greater climate-induced violence, European governments have in practice done relatively little to integrate climate change factors into their conflict prevention policies as these operate on the ground in specific conflict theatres. If anything after a rhetorical peak around 2009–10 the momentum behind the conflict–climate link has since then subsided. Constance Lever-Tracy insists little progress has been made and that governments, 'are still unprepared for the unexpected, unable to cope when it strikes, and have far too few resources available for recovery'.[2] After a decade of rhetorical commitment to climate security, ask even the most deeply involved and committed officials how much the EU actually spends on climate-related conflict resolution and clear answers are not forthcoming. A trawl through member states' conflict prevention-resolution programmes yields little in terms of initiatives related specifically to climate.

This is not to ignore those advances that have been made. Many claim that the EU has retained a lead role in the civilian capacities dedicated to conflict resolution. The focus on such civilian initiatives is seen as integral to the overarching cooperative approach to security. From 2007, the Commission-run Instrument for Stability (IfS) has widened the scope for conflict-related funding. Particularly notable increases have been forthcoming in disarmament, demobilisation and reintegration (DDR) funding and the EU has widened a programme for strengthening African capabilities for the prevention, management and resolution of conflicts. The EU now has at its disposal a plethora of disaster response bodies, including the Situation Centre, the Policy Planning and Early Warning Unit, GMES services, the Rapid Reaction Mechanism, the Monitoring and Information Centre, the European Space Agency Disaster Charter, a Continental Early Warning System, the EWS and the EU Satellite Centre.

On the back of the 2008 frame-setting Solana paper, consultations in the Commission urged the development of an 'environmental peacemaking' strategy. The on-going programmes of relevance to this are cited as including the EU Water Initiative and the EU Energy Initiative for Poverty Reduction and Sustainable Development, along with support for the GMES. However, a Swiss Peace report concluded that by 2011 little had been done to move the stated aim forward.[3] The EU's key strategy document on conflict prevention is the so-called Gothenburg Programme agreed in 2001. When EU leaders released a ten-year update of the Gothenburg programme their instructions of where conflict prevention needed upgrading made no mention of any climate-related factors.[4] In 2011 the European Peacebuilding Liaison Office (EPLO) carried out a review of this programme's first decade. It concluded that there was need for an updated strategy to reflect how conflict drivers had changed since 2001, as well as for improvement in still-weak implementation mechanisms. The European Security Strategy, with versions elaborated in 2003 and 2008, advanced on several key concepts, including human security, the role of civil society, gender issues and climate change as a cause of conflict. But the EU's well-known problems of inter-issue and inter-institutional coherence have not dramatically improved. This lack of coordination exists especially in relation to the poor linkages in practice between climate change, energy, trade and security decision-making.[5]

The IfS expanded from 139 million in 2007 to 282 million in 2011. Most of its funding has gone to short-term crisis response only 5 per cent to longer-term crisis preparedness. Much of its work is seen as related to climate factors: projects in the Horn of Africa to respond to tensions caused by drought and food price increases, by aiming to keep agricultural markets open and functioning; and in Bangladesh, building institutions to respond better to flooding. Climate conflict was not listed as one of the IfS's main priorities, but a new programme on 'Natural resources, post-conflict and post-disaster needs assessments' was a second-order issue of 'crisis preparedness'.[6] In 2011 a modest €1.5 million of its total spend was allocated for 'natural resources and conflict'.[7]

The IfS has funded some generic capacity-building related to disaster preparedness and supported United Nations (UN) efforts in this area too. However, it operates mainly in responsive mode. Plans are afoot to beef up its peace-building partnership as the main proactive instrument to tackle long-term issues driving climate conflict; this now includes support for civil society groups working on climate security to raise awareness among political elites. The Commission's humanitarian office, ECHO, has similarly been criticised for spending less than 10 per cent of its budget on preventative measures that would ensure better preparedness for climate-related disaster.[8] Officials insist that the post-2014 IfS is set to place more stress on climate-driven conflict. In 2013 the EU advanced with its €150 million Agadir Sahel programmes to strengthen resilience on climate change, directly linked to security interventions in Mali.

The EU has sought to enhance its early warning systems. It now has a whole plethora of institutions charged with ensuring timely preventive responses to crises. This is an area of strong development, although the risk exists of a rather inchoate system emerging. The European Commission Crisis Room, Regional Crisis Response Planning Officers, the Situation Centre, a European Rapid Alert System, the External Action Service (EAS) conflict unit and the EU Military Staff all have a role. Confusion reigns over the precise division of tasks, and officials admit that early warning responsibilities specifically linked to climate change are still not clear or easily operational. And significantly, no climate change-related factors are incorporated into the way that potential crises are monitored. The EU's Watch List is a motley collection of unstable or potentially unstable states that are of particular interest to one or more member states. It does not benefit from objective indicators for including on a regular basis encroaching situations induced primarily by climate change. The Central Intelligence Agency (CIA) has a unit for forward-warning climate crises; nothing so systematic has been created in European states.

By 2011, one of the stated aims following from the launch of the EU's new climate diplomacy strategy was to improve the Union's early warning mechanisms which were acknowledged still to be disconnected from the specifics of climate-induced conflict. SitCen officials insist that, as a concrete manifestation of climate security concerns, they now benefit from satellite tracking of the Arctic ice flows. However, satellite tracking of climate change instability in general was found in one extensive report of the Institute for Environmental Security to have been woefully limited.[9] The GMES initiative has been removed from the EU budget for the 2014–20 financial perspectives. The 2011 Climate Diplomacy initiative itself is still couched essentially at the level of recommending further regional studies and discussions before the EU adapts its conflict prevention strategies.

Officials in DG Climate concur that progress on the specifics of rendering conflict prevention programmes climate-sensitive has been woefully insufficient. While the 2010 floods in Pakistan 'woke people up', as these threatened political instability in a nuclear state, they admit that there is 'still a long way to go' in addressing climate conflict in a preventive fashion. Climate officials acknowledge that the focus has been on 'disaster response' much more than on 'disaster preparedness'. General common security and defence policy (CSDP) strategy still does not speak in the language of adaptation that mainstream climate specialists see as so urgent. The EU's 'comprehensive approach' to conflict that was developed during 2013 did not have any apparent operational consequences related to climate-induced instability. The EAS has coordinated joint sessions with African states and China on joint climate challenges. And militaries are on board and reputedly looking for concrete projects on the ground in the area of climate security. However, asked what tangible impact the agenda has had on conflict policies, DG Climate officials recognise that little operational change has been introduced beyond general pronouncements about the need to prepare more.

The British government's 2011 *Building Stability Overseas Strategy* provided the first cross-ministerial strategy for stability. Under its rubric, the UK committed to increasing the share of aid going to fragile states to one-third by 2014, with an overall increase in resources. One-third of UK overseas development assistance (ODA) must now help long-term capacity-building not just immediate poverty relief. The Stabilisation Unit and Civilian Stabilisation Group (with a roster of 1,000 civilian experts) were to be strengthened. A new Early Warning System was set up, managed by a dedicated steering board. A conflict Watchlist and Annual Horizon Scan would allow a more honed identification of risks, including through climate-related indicators, nourished by cross-departmental stabilisation and conflict assessments. A £20 million Early Action Facility would fund more pre-emptive capacity-building. In addition, a Defence Engagement Strategy would bring the Ministry of Defence (MoD) to work more on upstream prevention activities. And 'prevention partnerships' would be developed on all these issues with rising powers. This strategy document insisted that resources were already being spent in a way that reflected climate-linked conflict dynamics in the Horn of Africa, the Sahel and West Africa. Climate triggers were listed as one of the factors prompting the effort to upgrade and fine-tune conflict prevention efforts.[10]

The Danish government has created a new Peace and Stabilisation Fund which will make new funds available for climate security projects. Denmark has a long-standing approach based on increasing trust between involved parties and addresses problems related to the management of natural resources. The German MFA has explored the notion of track II initiatives to reflect on the interests of the groups most affected by climate change; the aim has been to build capacities to spend climate funds coherently and transparently with a special view to the governance challenges of fragile countries and regions, and a particular link to human security.[12] Germany now explicitly advocates 'conflict-sensitive adaptation practices', in particular linking track I and II approaches and using climate funds for governance challenges in fragile regions.[13]

There is a widespread consensus among policy-makers that the changes flowing from such formal initiatives are no more than embryonic. A 2012 independent assessment of the UK's conflict pool concluded that the basic policy had not changed; that nearly all funds were still spent on reacting to short-term crises; and that the MoD, Department for International Development (DfID) and FCO divided funds out for their traditional types of projects rather than working together on new joint approaches to underlying causes of instability. Moreover, by 2011 the DfID share of the pool was down to 5 per cent and the FCO share up to 60 per cent.[14] A similar stasis is evident elsewhere too. While Spanish ministers have made increasingly bold statements acknowledging the climate–conflict link,[15] the link between the defence minister and the secretary of state for climate change is relatively weak. Insiders acknowledge that this means that the UME's inception has not led to any change in the broader gamut of Spanish conflict prevention policies.

One team of experts has criticised European governments for not realising how far Middle Eastern democratisation will be hindered by climate-compounded resource constraints. A serious weakness of the EU's response to the Arab Spring is the absence of any effort to work with governments in the region to develop climate adaptability plans or to redirect meaningful amounts of aid towards addressing such priorities. Indeed, the risk has been that middle-income Middle Eastern states have fallen outside the rubric of large-scale climate finance, which has been channeled mainly to poorer states in sub-Saharan Africa – this despite the geopolitical and security implications being far more acute in the Arab world. European-funded initiatives in the region to date – Desertec and several other smaller plans – are judged to be of limited short-term help in boosting Arab countries' low carbon resilience.[16]

Organisations like International Alert argue that there remains a need in Western policies to understand the more detailed mechanisms through which climate change is likely to engender conflict. Extant policies rest on overly linear assumptions. The key variable is that where local civic organisations are weak, shared resource management becomes harder and a predatory state more easily takes command. Tensions between sedentary farmers and herding communities who are forced to move longer distances by climate stress need inclusive civic forums to resolve. Governance conditions are then a pivotal transmission belt between climatic conditions and instability. Non-governmental Organisations (NGOs) criticise European governments and the EU collectively for failing to incorporate such underlying drivers of instability into more climate-sensitive conflict prevention policies.

There is still no granular means of assessing the risk of climate-induced conflict and instability. Climate factors fed into early warning lists remain extremely broad-brush and not sufficiently detailed to understand how climate stress might actually trigger conflict. Moreover, generic warnings of droughts and floods are even less amenable to being moved from 'warning' to operational 'climate action.' Governments have struggled to incorporate climate-specific elements into their traditional conflict prevention programmes because they admit that more climate specialists would need to engage with the conflict agenda than has so far been the case.

European governments have stressed their desire to see the UN assume the lead role on the conflict prevention strand of climate security. But progress in the UN has been extremely limited. It has long been recognised at the highest levels in the UN that climate change is a factor of increasing pertinence to conflict prevention. As far back as 2001, then UN secretary general Kofi Annan declared that: 'Environmental degradation has the potential to destabilise already conflict-prone regions, especially when compounded by inequitable access or politicisation of access to scarce resources.' However, when the Swiss Peace Institute surveyed the UN's efforts to incorporate climate change factors into its conflict prevention programmes it concluded that 'the two issues of environment and conflict have not been explicitly linked in policy implementation'. The UN remains focused overwhelmingly on post-conflict crises rather than tempering the longer-term underlying causes of instability, such as climate stress. The UN Environment Programme's (UNEP's) specialised unit on Post-Conflict and Disaster Management Branch was merged into a wider Disasters and Conflicts Programme. European diplomats lament that tangible upgrading to UN policies on the ground has not been forthcoming.

While rhetorically EU member states have urged UN conflict prevention operations more systematically to lead on the climate–security link, in practice they have done little to help correct the UN's very limited role in this area to date. The UNDP's Bureau for Crisis Prevention and Recovery works on climate change, but mainly as a measure of risk assessment. It recognises climate change as a phenomenon that will accelerate current patterns of natural disaster, but does not explicitly deal with it as a feature of conflict prevention. There has been talk in the UN of a new environmental peacekeeping force – popularly dubbed the 'green helmets' – which could step into conflicts caused by ever-scarcer resources.[17] However, several European countries, including Germany, have argued that such a force is unnecessary as such functions are already being performed fully by the standard blue helmets, who have long intervened in areas beyond traditional conflicts. The Post-Conflict and Disaster Management Branch of the UNEP has begun working more systemically on climate change, migration and conflict in the Sahel.

The UN claims it has approached climate conflict from two directions: on the one hand, introducing environmental factors into the conflict prevention debate; on the other hand, introducing conflict prevention as a topic within the environmental debate. The first approach is reflected in discussions in the Security Council; the second approach is germane to the UNEP Conflicts programme. The two approaches are not mutually exclusive, but they *both* currently lack full implementation. The seminal 2001 Resolution 1366 on conflict resolution provides a working framework, but under this climate change has to be formally recognised as a 'root cause of armed conflict' by the Council before any action can be implemented. This has held back practical changes to conflict initiatives. Another problem is that the UN Global Environmental Facility (GEF) funds projects in a way that is linked to World Bank reform packages, which remain highly economistic and exclude any conflict prevention focus. While UNEP is one of GEF's implementing agencies, it has struggled to develop cross-agency climate-conflict programmes with other UN organs.[18] A subsequent 2011 Swiss Peace study found improvements were still piecemeal. In 2010 the UN Environmental Programme established environmentally driven conflict as one of its priorities, with an allocation of $50 million, along with a toolkit for UN country teams for how to address the climate–conflict link. But Swiss Peace concludes that it is still difficult to see any tangible upgrade or change to UN conflict prevention.[19]

A developmental approach to security?

The EU and European governments insist that they have engineered a tighter linkage between conflict prevention and general 'climate finance' – the aid ring fenced to help developing states with mitigation and adaptation. They have increased climate funding, and claim this helps mitigate tensions and conflict related to resource scarcity. There is a strong developmental orientation to such funding. This ostensibly reflects the broader EU 'soft security' approach: investment in economic and social development is seen as requisite to sustainable security. This logic appears to have been given particular resonance in the realm of climate security. This is certainly an evolving and strong aspect to European policies. However, the evidence suggests that the enormous range of climate-development funding forthcoming from European donors rather easily assumes its own security relevance, rather than this following from a careful or precise assessment of its impact.

The EU has developed initiatives to improve poor communities' access to energy. Policy-makers argue that this lends a distinctive dimension to conflict prevention and one that is centred more on mutual human security concerns rather than purely traditional state interests. If there is an identifiable area of climate-conflict specificity in EU policies it is that of 'energy access.' The EU has, since the mid-2000s, promoted its support for a wider access to energy as one of the most distinctive strands of its development cooperation, especially in sub-Saharan Africa but also increasingly in South Asia. Climate commissioner Connie Hedegaard insists that a focus on access to energy in ODA is and must continue to be the leading edge of the EU's linking of security and climate change policy.[20] Development commissioner Andris Piebalgs insists that he has taken an interest in climate change precisely because of the increasing share of ODA directed at ensuring a fairer and thus more harmonious distribution of energy resources. A new Commission communication in March 2013 promised to merge development and climate change issues into a single seamless anti-poverty policy.[21] Even fairly realist-tinged governments, like that led by the Conservative party in the UK, have insisted that climate change is a strategic problem primarily because it affects states that are already poor and weak, and have limited resilience.

European support has been forthcoming for the joint management of natural resources, such as river basin systems. The EU's line is that it seeks to use water disputes to 'build bridges' and temper tensions between local communities. This means tying into external agreements incentives to get third country partners cooperating with each other on joint management of scarce resources. Several member states fund initiatives on the joint management of water basins. The partnership for Climate Change and environment identifies the development of the Great Green Wall of the Sahara and the Sahel Initiative as priorities. These aim to address land degradation and desertification in the margins of the Sahara. A prefeasibility study commissioned by the European Commission and the African Union Commission has been used to catalyse member state financial support. One case study concurs that the collection of these initiatives indicate that in Africa the EU's entry point into climate security has been primarily through the mainstreaming of adaptation initiatives into holistic development aid programmes.[22]

A common developmentally oriented discourse is now more evident at the EU level and within member states. The 2012 European Development Report focused on how far EU development policy has progressed in addressing the nexus of inter-linkages between development, water use, sustainable energy and land use; this being the dimension of EU climate funding that has advanced most tangibly, and an area prioritised in the EU's 2011 Agenda for Action designed to upgrade and update development policies.[23] France's 2009 Climate Justice Plan, which attempted to close negotiating gaps prior to the Copenhagen summit, was predicated strongly on this development-oriented logic and professed to mobilise new aid allocations. The UK has moved to talk of 'resource security', reflecting a recognition that climate change may not directly cause conflict but rather puts stress on existing resource shortages; the policy agenda is more about the quality of political and economic governance.

In the Centre for Global Development's *2012 Commitment to Development Index*, EU donors received strikingly higher scores than other states for the quantity and quality of their climate-related aid.[24] For 2010–12 the EU allocation was €7.2 billion. This was split between 48 per cent for mitigation, 33 per cent for adaptation and 16 per cent for reducing emissions from deforestation and forest degradation. Just over half was given in the form of loans, and 48 per cent as grants. Nearly 60 per cent went through multilateral organisations.[25] By the start of 2013 the EU was on schedule with its fast start financing commitments. Proposals have been made to extend the 'blending' of Commission and member state aid fully into the area of climate security.

Denmark, Finland and Germany were the most generous proportionate contributors, allocating between 12 and 15 per cent of bilateral aid to climate projects. France and Spain have increased their role, up to around 10 per cent of their total aid going to climate adaptation. Sweden claimed that beyond its specifically defined climate aid it allocated more than $585 million in additional support to climate-related development assistance over the period 2009–11.The UK allocated a lower 3.5 per cent, despite running a large number of vanguard climate change programmes. It set up an International Climate Fund in 2010, an inter-departmental fund to channel the vast bulk of Britain's climate funding of £2.9 billion up to 2015; aid relevant to climate change is now running at just under 400 million per year, up from 100 million per year in 2005. In relative terms there was much that was impressive in these figures. Spain, for example, gave eight times more than the United States. In the summer of 2012 the Commission committed an additional €50 million. In October 2012 the EU agreed to provide China with €25 million worth of technical support to consider adopting some form of emission trading scheme.

Beyond the progress made in climate funding, however, three shortcomings persist in its role in climate security.

A tenuous link to security

First, critics say the EU's approach almost dampens the security logic through a discourse implying that the issue is little more than a need for more sustainable development. Diplomats admit in private that while access to energy projects may have had some useful impact, they have not been designed specifically with security aims in mind. So far, the commitment to climate security is difficult to detect within development programming. DfID business plans up to 2015 give little indication of how it might be given concrete substance. Indeed, DfiD insists on retaining 'firewalls' which prevent aid being spent on anything too security-related. DfID sees the climate security agenda as being about justifying its focus on the protection of vulnerable communities; DECC see it as justifying support for partners in multilateral climate talks; while the FCO sees in it money for security risks.[26] DfID officials admit that within the wide array of adaptation funding in developing states little progress has been made on 'climate-proofing peace-building'.

German diplomats acknowledge that, in practice, the conflict resolution dimension remains the weakest link in German approaches to climate security. Danish diplomats lament one example of what they see as the EU's inability to think and act geo-strategically in response to climate-induced social and economic changes: while China is moving massively to take advantage of newly accessible rare earths in Greenland and offering comprehensive economic partnership in consequence, the EU is merely offering ineffectual 'soft' and off-the-shelf development projects to help people deal with the micro-level socio-economic adjustment required by climate change.

The vanguard work of influential peace-building organisation International Alert admonishes European donors for failing to close the disconnects between the conflict resolution, development and climate adaptation policy-making communities. Security is still 'the elephant in the room' within international climate talks and the plethora of climate diplomacy forums that now exist. Donors still need fully to incorporate 'conflict sensitivity' into their adaptation funding. The key is to help make societies more resilient to climate variability. This reinforces the need to strengthen local economic, social and political capacities in developing states. The inter-linkages between climate and conflict interventions remain poorly understood and articulated. Many adaptation projects inadvertently make societies less resilient. This is because rents from new technologies are captured by elites in a way that compounds governance pathologies. Governments conceive adaptation too narrowly as a matter of protective, physical infrastructure or renewable projects; in fact, it is a matter of improving governance to make political elites more responsive. Most climate financing gives recipient governments full freedom to spend the resources as they see fit, which International Alert judges problematic in terms of pressing those same governments towards political reform. The huge amount of climate financing now provided fails to incorporate any tangible political dimension. In turn, peace-building projects are still not 'climate-proofed'; donors support job creation and economic development in fragile states, aiming to reintegrate combatants back into existing modes of production – without recognising that these modes are unsustainable from an environmental perspective and thus risk storing up greater instability over the long term.[27]

The second *Climate Vulnerability Monitor* published at the end of 2012 criticises Western donors for uncritically assuming that investment in generic processes of economic development reduces the need for adaptation, which still represents only 14 per cent of climate funding to the least developed states. This leads them to be so short-sighted that in many instances they spend more in disaster response than the investment in adaptation-defences that would have been needed to avert the emergency.[28] UK officials admit that countries like Bangladesh judge that they have got little benefit in terms of different types of climate aid projects. European development banks recognise the limits to their climate-proofing of loans made to emerging and developing economies.

Policy-makers admit that they are already struggling to design governance support sufficiently tailored to local conflict dynamics: it has proven too much to require another level of bespoke design, now tailored more specifically to the conflict-generating dynamics caused by climate stress. So far there is little evidence of much attention being given to the way that the renewables sector is managed, or any attempt to link governance initiatives to renewables policy. European policy statements have certainly presented good governance in renewables as essential for energy security. Policy-makers acknowledge that simply subsidising renewables in developing countries is unlikely to produce maximum benefit without reform to governance structures in those third states. But in practice, little tangible action has been forthcoming. To date, there are few signs of adaptation initiatives being linked to the EU's myriad projects in the area of democratic governance, for instance in a way that would enhance decentralised participatory programmes that empower communities in climate management.

Limited amounts

Second, economists point to the short-fall in existing climate finance: current allocations stand at around $15 billion a year; the 2020 target is $100 billion a year; $200 billion is required to make any kind of tangible impact.[29] The amount of climate funding negotiated through the UNFCC is a tiny percentage of the potential estimated cost of climate change: 0.5 per cent compared with 20 per cent of Organisation for Economic Co-operation and Development (OECD) gross national product.[30] Yet still governments have haggled over the distribution of such funding allocations. While adaptation is now being defined in a broader sense than mere physical defences – to include the need for changes in development models and early warning – in 2011 developing states received only $960 million for adaptation, compared with the $8 billion a year needed by 2015.[31] Development commissioner Andris Piebalgs acknowledges that the scale of climate financing agreed so far within mainstream Commission development budgets is 'meaningless' as a security instrument. He argues that the amount of funding offered can assist in a minor amount of adaptation at the margins, but is totally inadequate in helping prepare more anticipatory solutions to the strategic impact of global warming. DG Energy officials recognise there is still a need for more systematic use of research and development budgets to include the EU's neighbours in renewables development. Member state officials also acknowledge that dialogue with consumer countries on cooperation in climate aid projects has so far remained more limited than what is needed to have any impact on international relations.

The EU raised hackles on the first day of the December 2010 Canucn summit by revealing that half its fast start funding would take the form of loans and private equity instead of grants. Many environmental campaigners express concerns that the EU still has to demonstrate that it will resist the temptation simply to divert existing development aid or to use the climate adaptation agenda as a covert means of introducing new forms of conditionality and even trade barriers. Governments have rejected novel proposals, such as that to ring-fence future taxes on banks for climate adaptation. Italy was singled out for a particularly poor record in actually delivering on its promised funding.[32] Member states still have different ideas on additionality and on reporting criteria. Most governments over-report climate funds to least developing countries. France is most suspect on the issue of additionality – much of its climate funding commitment simply repackages existing aid projects in disingenuous fashion.[33] Member states have pushed up to 50 per cent the share of ETS revenues to be allocated to climate projects, so as to reduce the burden on their own budgets; this represents another dent in the spirit that new climate aid promises should bring additional money to the table, not simply divert resources from other revenue sources.

The 2012 'Rio plus 20' summit was a virtual non-event: the EU kept a low profile and did not make commitments on the link between sustainable development and climate change. Diplomats and even the EU climate commissioner herself lamented that this was a lost opportunity of merely rhetorical re-statements of basic principles. Some politicians and officials in some member states argue that the EU should reduce subsidies for renewable energy and adaptation outside Europe and that funds channelled instead towards internal energy efficiency represent a much better investment for security. Poorer member states now express explicit opposition to huge transfers for climate funding to the likes of China, countries growing fast and suffering less economically than many EU governments. There are concerns that EU funding for CCS in China is being wound down without any real impact having been made.[34]

Clash of priorities

A third weakness is that European governments have not acted in ways fully consistent with their proclaimed approach to conflict and climate change. In practice, significant tensions have surfaced between different policy aims. Often conflict is not simply driven by local groups and tribes battling for scarce resources but the exploitation of those resources by European companies. Inconsistencies between policies abound: the EU over-fished off Somalia and choked off local exports, then, as local fishermen turned to piracy, had to pick up the pieces with the Atalanta mission. While the development and conflict prevention goal has been to increase access to electricity grids for poor communities, policy-makers charged with registering progress on climate change goals complain that this sits uneasily with the need to reduce emissions. They worry that the EU risks pursuing an approach to resource-conflict that actually renders such conflict more likely in the long term. Oxfam and other critics slammed the Energy 2020 paper for undermining the principle of local access to energy resources; the document explicitly suggested the EU's priority should be to secure access to African resources beyond hydrocarbons for Europe's own energy needs. Critics complain that European donors have not used aid resources sufficiently to encourage energy decentralisation, for example by buying generators for villages to cut down on the need for vulnerable cross-border infrastructures.[35]

A frequently expressed admonishment heard from environment ministries is that 'climate security' has become a label under which development and climate aid resources are being deployed in the service of short-term European strategic self-interest rather than genuinely global progress on climate targets. European donors have funded projects under the Clean Development Mechanism (CDM) and other mechanisms designed to give a quick reduction in their own emissions figures rather than of a kind that enhance local access to renewable energy. Moreover, they have done relatively little to assist the 'losers' from new renewable projects, as a shift in energy patterns nourishes local grievances in many poor countries.[36] The CDM is largely disconnected from local social dynamics, simply implanting technology that often does not organically dovetail with local conditions. It has served simply as a crude transfer mechanism: nearly all buying has been from EU states; nearly all selling has been to China. Some experts criticise governments for thinking that adaptation is a matter of spending funds to help poor farmers protect their existing way of life, when they need a far more forward-looking vision that helps threatened communities into urban jobs.[37]

Developing countries have been critical of European donors for shifting too far away from mitigation towards a very generalised focus on adaptation that reduces obligations linked to their own pollution. With efforts to limit global warming to tolerable levels (mitigation) no longer sufficient, more experts advocate at least some complementary focus on how to manage its effects (adaptation). EU officials indeed insist that the focus on mitigation projects, aimed at meeting quite arbitrary and unrealistic emissions targets, has shifted out of strategic self-interest to focus more on adaptation – that is, on helping the most vulnerable countries contain and temper the impact of climate-induced damage. They lament that spending on mitigation is still easier to sell politically; building wind farms is a very visible and tangible action. This contrasts with the more nebulous political and social measures that are necessary for civic adaptation to impending climate insecurity. Yet, some scientists argue that climate security leads to an artificial separating out of those extreme weather events caused by man-made climate change and skews aid for adaptation towards cases driven by a climate security agenda and away from more general vulnerability of the poorest communities. Many in developing states complain that this type of thinking reflects an effort on the part of richer states to avoid any legal penalties for climate disasters that 'their' pollution causes in poor societies. The tension between security and development objectives permeates ever-sharper debates over the balance between mitigation and adaptation.

Conflict and migration

As already pointed out, the relationship between conflict and climate-triggered migration is a staple of much analytical prediction. Rhetorically, European ministers most commonly repudiate the 'fortress Europe' approach to managing climate migration. The stated preference is for a more subtle approach based on cooperation with third countries aimed at 'managing' the impact of climate change on migratory flows. The then British energy and climate change minister Chris Huhne insisted that UK policy aims to avoid the lifeboat-mentality: 'faced with changing patterns of migration, the international community cannot simply pull down the shutters. We must have the political will and institutional readiness to cope with greater numbers of people fleeing environmental change … many countries will be tempted to raise their defences against fleeing people. Barriers that will undermine a liberal world order that has served Britain well.'[38] Migration is talked of as a positive adaptation strategy rather than something only to be contained. One European official is adamant: closing borders would be 'the worst possible response'.

In practice, European policy on climate and migration remains ambivalent. It is well-known that EU member states have gradually tightened rules on migration into Europe. The formal EU policy has been to elaborate a balanced policy through its so-called Global Approach to Migration. This involves non-EU states agreeing to cooperate on stemming flows of irregular migrants, while in return the EU frees up the flow of legal migration. In general, the figures suggest a tightening of entry conditions. In the period 2007–11 immigration into Europe declined by 6 per cent, while emigration from the EU increased by 13 per cent. Asylum applications have fallen by a third since the early 1990s; acceptance rates have tumbled to only 29 per cent. Member states still treat claims differently, despite attempts since 1999 and the Dublin regulation to agree a common EU asylum system; this was supposed to be completed in 2012.[39] In May 2011 a Commission communication advocated a new series of dialogues on this balanced approach. Commitments were made for increasing the powers of Frontex, the border control agency; appointing an EU anti-trafficking coordinator; a stricter returns policy; and an entry–exit control system.[40] Some member states have pressed for a fully fledged EU border guard. The European border surveillance system (Eurosur) proposed by the Commission in 2008 and the 'border fund' for building up border guards are now being beefed up dramatically. After the Tunisian revolts, member states changed Schengen rules to make it easier to reintroduce national controls.

The influence of climate-induced migration in explaining these policy trends has been negligible. Indeed, it is striking how absent climate concerns have been from the evolution of European migration policies. The aforementioned October 2012 EP report worried that EU institutions had failed to devise contingency plans for an increase in climate-driven migration.[41] Former minister Charles Clarke laments that there is still no linkage of migration aims with foreign, security, trade and aid policies.[42] Migration is one of the areas of climate security most exhaustively covered by analysis, but least incorporated into policy-making. In private, officials from DG Home (the Commission's interior ministry equivalent) admit they have expressly eschewed any involvement in the climate security agenda. Their view is that the general provisions for strengthening border controls and emergency surveillance mechanisms will be appropriate to and increasingly available for climate-induced migration. Migration officials see the climate security agenda as unduly complicating the picture when even the current drivers of mass movements are proving so hard to manage.

The EU still has no mapping of what migration flows are likely from different parts of the world as a result of climate change. It lags behind the US in preparing for climate migration. US government bodies have begun running exercises focusing on the impact of displacements within other regions, for example how flooding in Bangladesh will drive migrants into India, causing instability which in turn impacts on Western interests. European organisations have still not included migration in their climate risk mappings because of the uncertainties over its scale and nature; and this issue is still not part of inter-regional negotiations, say, between the EU and African Union.[43] A small number of more climate-specific initiatives are now in evidence. Member states agreed to set up a European Emergency Response Centre, in part with extreme weather emergencies in mind. The EU has funded the Norwegian-Swiss Nansen process, which in 2012 grew out of a conference that drew up a series of basic principles for managing displacements and aimed at deepening cooperation between governments on this question. However, these measures are few in number.

The most exhaustive report prepared under the auspices of the British government's Foresight centre declines to examine existing policy responses, as it paints general scenarios of different types of migratory flow. Its main message is implicitly critical of the approaches adopted so far. The report's message is that policy needs to attack the root causes of migration but cannot hope fully to contain or prevent increased flows. Therefore more strategic planning is required to manage the flows that will inevitably arise. The challenge is to make sure operational challenges do not become geopolitical challenges associated with mass, unplanned, unpredictable and illegal movements across borders. Governments err in seeing climate migration only as a problem rather than a positive step of 'transformational adaptation' that can release human and developmental potential; they have not funded preparedness in this sense from the Green Climate Fund or other adaptation initiatives.[44]

The so-called Stockholm programme agreed in 2009 as the EU's main strategy for internal security calls for greater focus on climate change as a driver of security-relevant migratory flows. The Commission's €180 million 2011–13 strategy paper that guides funds under its thematic programme for 'Cooperation with third countries in the areas of migration and asylum' explicitly commits to working more on the nexus between climate change and migration. This is indeed listed as one of its priority objectives. The paper commits the EU to increasing funding across a range of relevant instruments, including not only the migration-specific thematic programme but also the Instrument for Stability and various environmental programmes, and to ensuring coherence between these various initiatives. However, the Stockholm strategy essentially advocates further research on the relationship between climate change and migration. The nine lines it devotes to this topic (out of twenty-five pages) contain no substantively operational commitments.[45] Similarly, the 2012 Global Approach migration policy briefly mentions climate factors, but offers no specific policy remedy or commitments.

In April 2013 the EU published a new staff working document on climate change and migration. Significantly, this was prepared by the development and cooperation agency, DevCo, rather than DG Home. In line with most recent research, it downplays the likelihood of mass flows into Europe itself resulting directly from climate stress in developing states. Rather it places most stress on movements within developing countries themselves and the problem of internally displaced people. The focus is on development-related resilience-building. It states that the core policy aim should be to improve development opportunities to allow people to stay put in local communities as resources became scarcer. At the same time, the Commission acknowledges that more needs to be done positively to assist relocation where this would help access to resources. It admits that so far no more than a few limited projects have been funded to give substance to this more positive approach and that recipients' country strategy papers still include no mention of the climate–migration link that would serve as a basis for practical aid programming decisions.[46]

Two authoritative migration experts welcomed the new development-oriented focus and the commitment to move beyond ad hoc and defensive reactions to individual disasters, but observed that most policy ideas contained in the new documents are still aimed at reducing migration rather than promoting relocation as a positive form of adaptation.[47] Another expert opines that EU member states have so far been reluctant to radically deepen EU-level cooperation in response to the prospect of climate-induced migration, indeed arguably the latter has made them more nervous to surrender sovereignty in this area.[48] Most diplomats do now argue that the focus needs to be on intra-regional flows not flows into Europe. They tend to argue that climate migration may be of more indirect than direct concern to Europe: displacements from one developing state to another, or from one region to another within the same state, may trigger conflict and instability that then affect Western interests. Several new cooperation programmes in consequence focus on this dimension of climate migration – although funds remain limited compared with those pumped into standard border controls.

A running debate exists over the formal status of 'environmental refugees'. The difficulty has been in legally defining an 'environmental refugee', as climate-induced displacements are not likely to be driven by identifiable perpetrators to the same extent as with political refugees. Legal experts insist it would be difficult to draw the line between a 'climate refugee' and a 'climate migrant' who takes a pre-emptive decision to move in order to avoid a gradually deteriorating environment at a moment when a genuine choice still exists in that decision. Moreover, as commented in chapter two, it may be impossible to separate out climate from general socio-economic factors in determining the reasons for migration. It may in most cases be difficult to distinguish a specifically climate migration from other forms of migration. The UN convention on refugees was crafted in 1951 to deal with those fleeing from totalitarian regimes; it is looking out of touch with the very different factors driving refugee flows today.

EU member states have not supported the idea of including a new category of migrants in the EU's international agreements or policy frameworks, as many developing countries urge. A systematic stakeholder consultation on the Global Approach to Migration carried out in 2010–11 revealed that over 80 per cent of respondents favoured strengthened policies aimed at climate-induced migration and revealed that member states were not supportive of a new legal category of environmental refugee, wary of recognising any new obligation to accept such migrants.[49] Governments argue that if climate migrants were given the status of refugees this could prejudice the reception and resources given to those fleeing really acute political violence as a matter of absolute urgency. The aim is more modestly to get climate factors incorporated into international rules on internally displaced persons. DG Home has suggested a status of 'permanently forced migration' as a new category to get around the refugee problem. The April 2013 working document makes no commitment to consider any new legal status for climate-related movements; indeed, it argues there is no need for 'refugee-type protection' specifically on climate-related grounds.[50]

Many Green MEPs have been pushing for a decade for the EU to recognise climate refugees; the breadth of this campaign in the EP has expanded to other parties. Many meetings involving several hundred NGOs have been held in the parliament. Several Scandinavian states and the Belgian senate have considered how protection might be enhanced for those escaping climate-related events on a pragmatic basis. However, Finland is currently the only member state with a temporary protection provision that specifically offers help for those fleeing from 'environmental catastrophe' – although in practice, this has not been used. The EU Temporary Protection Directive does not explicitly offer help related to climate-induced disasters. In short, moves opening the scope for accepting climate-related migration remain scattered, tentative and ad hoc.

Conclusion

Climate change has entered the lexicon of conflict prevention. As yet it has not produced any major strengthening of European conflict prevention programmes, in either quantitative or qualitative terms. Climate security is mostly judged to be a generic part of conflict policies; that is, very broad efforts to increase the resilience of developing societies to social and political tensions are also taken to constitute preparedness for the more intense effects of climate change. Some more climate-specific dimensions have gradually been added to the range of EU conflict prevention initiatives. However, these are not yet of a magnitude to suggest that a high priority is attached to this area of policy or that 'climate conflict' is perceived as a particularly urgent risk. Under- rather than heavily securitised over-commitment appears to be the most salient feature in European conflict prevention units.

Doubts also persist over the way in which 'access to energy' is understood; sceptics fear that well-meaning initiatives aimed at tempering 'energy poverty' may worsen societal tensions over the longer term. The relationship between development and security aims is still disputed among EU policy-makers. For many, the security-development linkage still needs to be inverted. The finer detail of European climate funding reveals that in qualitative terms continuity prevails over the adoption of new approaches. Much new funding is devoted to very traditional projects of long pedigree and part of a broadly defined environmental agenda more than they are targeted systematically at the now-emerging political and geo-strategic spill-over of climate change. The risk is of governance-driven conflict, compounded by climate change, rather than of directly climate-driven conflict per se; but this is not how the EU has approached the challenge in any really significant manner. Indeed, the danger exists that the way in which the EU is spending its climate aid might inadvertently be worsening security dangers. Developing state diplomats still express fear that despite all the EU's rhetorical stress on civilian conflict prevention policies, climate security is a code for more hard security interests – this is despite the fact that the EU is hardly straining at the bit to send troops abroad to 'deal with climate change'. Restrictive measures on migration fit into this pattern, although the lack of linkage between climate security concerns and this area of policy remain striking. Migration is already highly securitised, so any additional impact from climate planning has been modest. In sum, the relative paucity of climate-specific conflict prevention funding suggests some fundamental choices still need to be made: whether, in security terms, it is more important to focus resources on influencing the major emitters or on conflict-prone societies.

Notes

1 N. Mabey, 'Climate change and global governance', E3G memo, October 2009, www.e3g.org, p. 10.
2 C. Lever-Tracy, *Confronting Climate Change*, London, Routledge, 2011.
3 Swiss Peace, *Environmental Peacebuilding: Managing Natural Resource Conflicts in a Changing World*, Swiss Peace report, 2011, p. 68.
4 European Council, *Council Conclusions on Conflict Prevention*, COSDP 611, June 2011.
5 www.eplo.org/assets/files/2.%20Activities/Working%20Groups/EEAS/EPLO_Review_Gothenburg_Programme.pdf.
6 European Commission, *2011 Annual Report on the Instrument for Stability*, COM (2012) 405, 2012; plus for details on programmes, the four volumes of the accompanying Staff Working Document.
7 www.eeas.europa.eu/ifs/docs/c_2011_4451_en.pdf.
8 http://ec.europa.eu/echo/files/media/publications/annual_report/annual_report_2010.pdf.

9 G. de Ville and R. Kingham, *Recent Trends in European Union External Action in the Fields of Climate, Environment, Development and Security*, The Hague, Institute for Environmental Security, 2011.
10 Department for International Development, the Foreign and Commonwealth Office and the Ministry of Defence, *Building Stability Overseas Strategy*, London, The Stationery Office, July 2011.
12 German Ministry of Foreign Affairs, *Climate Diplomacy in Perspective: From Early Warning to Early Action*, Berlin, German Federal Foreign Office, October 2011.
13 German Federal Foreign Office, *Climate Diplomacy: Reducing Risks for Security*, Berlin, German Federal Foreign Office, 2012, p. 17.
14 Independent Commission for Aid Impact, *Evaluation of the Inter-Departmental Conflict Pool*, report 12, 2012.
15 www.defensa.gob.es/gabinete/notasPrensa/2011/02/DGC_110215_ministra_presentacion_cuadernosIEEE_CESEDEN.html.
16 N. Mabey, S. Schulz, T. Dimsdale, L. Bergamaschi and A. Amin, *Underpinning the MENA Democratic Transition: Delivering Climate, Energy and Resource Security*, London, E3G, February 2013.
17 www.euractiv.com/en/climate-environment/un-security-council-consider-climate-change-peacekeeping-news-506690.
18 Swiss Peace, *Linking Environment and Conflict Prevention: The Role of the United Nations*, CSS, Zurich, 2008.
19 Swiss Peace, *Environmental*, p. 47.
20 Carnegie Endowment for International Peace, *Climate Security and Green Technology: An EU-US Perspective*, transcript of conference, 8 April 2011.
21 European Commission, *A Decent Life for All: Ending Poverty and Giving the World a Sustainable Future*, COM (2013) 92 final.
22 R. Rodgrigues de Brito,' The securitisation of climate change in the European Union', in OECD, *Global Security Risks and West Africa – Development Challenges*, Paris, OECD, 2012.
23 European Development Report 2012, *Facing Scarcity: Managing Water, Energy and Land for Inclusive and Sustainable Growth*, Brussels, Europe Aid, 2012.
24 Centre for Global Development, *2012 Commitment to Development Index*, Washington, Centre for Global Development, 2012, p. 3.
25 http://ec.europa.eu/clima/publications/docs/spf_startfinance_en.pdf.
26 K. Harris, 'Climate change in UK security policy: Implications for development assistance?', ODI working paper 342, 2012, p. 13.
27 D. Smith and J. Vivekananda, *Climate Change, Conflict and Fragility*, London, International Alert, 2009, p. 4, p. 8, p. 11 and p. 22.
28 Climate Vulnerable Forum–DARA, *2nd Climate Vulnerability Monitor*, Madrid, DARA, 2012, p. 20 and p. 25.
29 E. Haites, 'Climate change finance', special edition of *Climate Policy*, 11/3, 2011, 963–69, p. 967.
30 Mabey, 'Climate change', pp. 5–6.
31 O. Hoffermann, 'Adapting to a warmer world: No going back', *Nature*, 491, 28 November 2012, 659–61.
32 www.europeanvoice.com/article/2010/11/italy-blamed-for-eu-failure-on-climate-change-aid/69421.aspx.
33 I. Scholz, 'European climate and development financing before Cancún', EDC2020 Opinion 7, December 2010, p. 2.

34 T. Burke and N. Mabey, *Europe in the World*, London, E3G, 2011, p. 30.
35 Citpax, *Cambio Climático y Seguridad Global*, Madrid, Citpax, 2011, p. 11.
36 European Report on Development 2012, *Facing Scarcity*, p. 108 and p. 122.
37 P. Collier, *Plundered Planet: How to Reconcile Prosperity with Nature*, London, Allen Lane, 2010.
38 C. Huhne, 'The geopolitics of climate change', speech at the Royal United Services Institute, 7 July 2011.
39 C. Clarke, *The EU and Migration: A Call for Action*, Centre for European Reform Essay, November 2011, p. 9 and pp. 14–15.
40 European Commission, *A Dialogue for Migration, Mobility and Security with the Southern Mediterranean Countries*, Brussels, COM (2011) 292, 2011; European Commission, *Communication on Migration*, COM (2011) 248, 2011.
41 Ibid., p. 9.
42 Clarke, *The EU and Migration*, p. 29.
43 M. Werz and L. Conley, *Climate Change, Migration and Conflict*, Washington, Centre for American Progress, January 2012.
44 Foresight, *Migration and Global Environmental Change, Final Report*, London, The Government Office for Science, 2011, chapter eight.
45 European Commission, 'Cooperation with third countries in the areas of migration and asylum', 2011–13 multi-annual strategy paper, especially pp. 27–28.
46 European Commission, *Climate Change, Environmental Degradation and Migration*, staff working document, 8556/13, Brussels, European Commission, 2013, citations from p. 22, pp. 28–29, p. 34.
47 A. Geddes and W. Somerville, 'Migration and environmental change: Assessing the developing European approach', policy brief 2, Migration Policy Institute Europe, May 2013.
48 S. Srichandan, *Climate Change, Migration and Security*, November 2009, www.chathanhouse.org.uk.
49 European Commission, *Summary of Stakeholder Responses to the Public Consultation on the Global Approach to Migration*, Brussels, European Commission, July 2011, pp. 7–8.
50 European Commission, *Climate Change, Environmental Degradation*, p. 18.

8

THE GEO-ECONOMICS OF CLIMATE SECURITY

A crucial set of strategic considerations derives from the questions of who has control over low carbon technology and the kind of rules that govern its internationalisation. An incipient geo-economics of climate change will be of inordinate importance in reshaping the international system. What form will that geo-economic dynamic take? Will globalisation deepen, tying erstwhile fractious rivals together in networks of cooperative and mutually constraining interdependence? Or will governments seek more mercantilistic autarchy as the preferred means of tempering the uncertainties of climate change? How will the international development and distribution of renewables technology be managed? How will the deployment of geo-engineering solutions be controlled and decided? Is security best guaranteed by cooperation in low carbon technologies, or do European governments perceive a need to temper markets in order to retain control and reduce dependence on external sources of renewables?

In exploring these questions, this chapter shows that climate change is now approached as a core pillar of European economic security. In some ways, this geo-economic dimension has filtered into European Union (EU) policy responses to a greater degree than have hard security aspects. The feedback of cause and effect increasingly intertwine: governments' attempts to manage climate change have pushed them to modify geo-economic preferences; the latter are in turn one of the factors that act increasingly strongly upon climate security. The EU collectively and member states through their national foreign economic policies have inched towards a geo-economics of climate change. However, a fundamental tension remains evident: freer markets are seen as necessary to get scarcer resources to where they are needed and to facilitate the spread of renewables; but globalisation is simultaneously seen as sustaining the economic model that lies at the very root of global warming. Emerging EU policy mixes support for open markets and government-backed commercial diplomacy. The European response may be interpreted as seeking to strike a balance between interdependence and autarchy as the driving logic of security. Yet it is doubtful that policy is sufficiently attuned to the widest economic aspects of climate security as opposed to low carbon commercial gain.

Geo-economic dilemmas

In recent years, a plethora of concerns has surfaced over climate-related economic security. A 2013 Chatham House report observes a general trend towards nationalistic, antagonistic and protectionist, cartel-like responses to resource insecurity – the antithesis of a focus on resource resilience across the most vulnerable countries, upon which global security in fact depends.[1] In 2011 floods in Thailand caused Honda to delay the launch of a new car in the UK. Tesco's publicity stunt of removing all products from its shelves requiring interlinked global supply chains – which left a virtually empty shop – raised the profile of such concerns further. Insurance companies say that the whole logic of the insurance sector will have to be rethought; disaster-related liabilities will become impossible to meet and premiums could become prohibitively expensive, which could threaten a vital underpinning of global commerce. Access to rare earths has become a priority issue for green technology, increasing dependence on China, Brazil, Russia and some states in central Africa. Over 90 per cent of global rare earths production is from China, which will need more and more of this for its own domestic use rather than exports. Climate impacts are starting to feed into some very elemental realpolitik. China and Taiwan have competed fiercely for influence in the Pacific Islands, each offering to support the latter's concerns over rising sea levels and provide highly lucrative commercial deals in return for their support on the question of Taiwan's recognition. As a result, France complains that it has lost influence in the region. China is also striking deals with Russia over food imports and beginning to include clauses on food supplies in its deals with developing states, requiring stipulated quantities of supplies to be shipped to China as a condition for the funding of infrastructure development. Even the Group of Twenty (G20) has moved towards carving out a role for itself on resource security.

Many analysts argue that future economic security can only be achieved through a fundamental change to the current market-growth based economic model, with a focus on resource productivity rather than growth rates.[2] They insist that the climate challenge cannot be conceived as 'others' needing to 'adapt' more than the West needs to examine its own basic economic model. The 'south' has been ambivalent over the climate security agenda because it is painted as the source of security problems for the West. The post-2008 economic crisis has been a diversion as it has turned European policy-makers' attention back on returning to growth by any means possible. Planning for qualitative change to the economic model has been relegated in importance. The plethora of 'green growth' investments highlighted as part of stimulus packages is still based on increasing consumption. Critics insist it is not possible to have growth without impacting the climate, and that moreover climate change itself will reduce growth potential. From this perspective, Europe's crisis is not of the euro per se but of an unsustainable model of accumulation.[3] Dieter Helm argues that dealing with the climate challenge must involve lower rates of growth and qualitative change, not quick-fix returns to growth through old-style Keynesian infrastructure projects. A carefully planned shift to low carbon will take a long time and not generate significant levels of employment or growth immediately. To the extent that the crisis was the result of over-consumption, it should engender demand for a new green model. Yet politicians have debated the crisis only in terms of a binary choice between austerity and a quick return to consumption-fuelled growth.[4]

Simon Dalby makes the point that climate security is less a matter of 'rural squabbles' over resources that might tip into localised conflict than it is of the whole underpinnings of the global economy – something that as of mid-2013, he laments, was still not taken into account by international policies. Indeed, the increasing focus on 'land grabs' to hoard future resources risked igniting broader instability, because of governments' unduly narrow understanding of both adaptation and environmental security.[5] Given increasing geo-economic competition for resources, some analysts argue that the West's support for market mechanisms will damage its own security, as large quantities of exports will not find their way onto the open market. Climate change will shift alliances in a way that will hasten Europe's international geo-economic marginalisation.[6] Writings on the post-crisis global economic system suggest that zero-sum dynamics must prevail primarily because of the future scarcity of finite resources. The West gained control of 'enabling resources' before it committed to liberal multilateralism; rising powers cannot be expected to follow a different route and will fight hard for direct control over diminishing natural resources. This means that the West's relative decline could easily tip over into an absolute impoverishment and strategic vulnerability.[7] An autumn 2013 Citi Bank report observes a new 'energy Darwinism' as competitive support for different technologies has intensified.[8]

Others insist that precisely the need for innovation in the economic model renders open and dynamic markets even more essential for economic security. James Martin argues that the need for innovation in a context of future adaptive complexity renders central planning unworkable and undesirable; governments err in thinking climate change requires more dirigisme when in fact it needs quite the opposite, more decentralisation to small firms operating in dynamic market contexts.[9]

Debates have sharpened over the policy implications of these concerns. Less conspicuous than the ostensibly more dramatic issues of climate conflict and migration, the geo-economic dimensions of climate change are beginning to attract attention. Many European officials and diplomats stress that they see climate security primarily as an issue of economic security. A core part of the flagship EU2020 Strategy, aimed at boosting long-term European competitiveness, was a formalised strategy committing the Union to mainstreaming resource scarcity within EU economic policies.[10] A United Nations Environment Programme (UNEP) report observed that European trade ministries have dramatically increased their engagement over climate issues, supporting a range of new studies and seminars.[11] In October 2012 the Commission presented an update of EU industrial policy that committed to strengthen the use of reciprocity to create more level playing fields for European investors overseas and to back this up with a 'raw materials diplomacy' to secure access to vital supplies.[12]

Diplomats highlight the importance of a new EU narrative being prepared in 2013 to link climate security more to opportunities for green growth, as a means to re-inject momentum on the back of the eurozone crisis and to get sceptical economics and trade ministries on board. The new geo-economic narrative aimed at promoting a more positive public diplomacy strategy. A shared feeling across member states is that China has stolen a lead in these geo-economic strands. Germany has been particularly keen on giving a more geo-economic edge to policy, with the country widely seen as the most successful European geo-economic power.

As energy and climate minister Chris Huhne argued that the first and most vital aspect of climate security will be its effect on trade.[13] In the UK a 'resource scarcity working group' has been set up and the government launched a new 'action plan for resource security' in March 2012. This defined climate change as one aspect in a cluster of issues threatening resource security and requiring a geopolitical response; access to speciality metals on the road towards low carbon was targeted as a particular concern. A range of partnerships was set up between government and business and a plethora of funding initiatives was launched; the approach in strategic terms was defined as 'business resilience'.[14] Defending the 'circular economy' is the new leitmotif of FCO thinking on climate security: even if the UK is not directly affected by climate conflict it will be hurt by any interruption to imports and global supply chains. UK diplomats also argue that it is the geo-economic dimension that resonates with an otherwise climate-sceptical public opinion. Certainly, the most senior of officials in DG Energy reveal that there are already incipient concerns over a concentration of renewables supplies.

Like other areas of climate security policy, however, this strand of deliberation remains embryonic. One former minister admits that the EU has so far made little progress on thinking what rules will be necessary to govern the distribution of new technologies in a way that prevents the emergence of new geopolitical tensions. The rules governing international trade in low carbon technology remain strikingly undeveloped. The relationship remains deeply contested between free trade, on the one hand, and the nurturing of green industries, on the other hand. Two inter-related issues have attracted fierce debate. First, how far 'green trade' should be given preferential treatment within international deals. Second, the extent to which domestic subsidies for green energy and technology should be allowed. At the end of 2012, the US National Intelligence Council warned that a key transmitter of climate tension is the persistent lack of an international framework for dealing with export controls on food exports.[15]

Opposing effects exist. On the one hand, trade generates economic growth, which is closely linked with increased energy use. On the other hand, open trade should increase the availability and lower the cost of climate-friendly goods and services; moreover, an increase in income levels brought about by free trade is likely to lead the public to demand higher environmental standards. The World Trade Organization (WTO) and UN have argued that the balance between these negative and positive effects depends on a country's level of development. Researchers have suggested an Environmental Kuznets Curve. For countries with low levels of per capita income, increased economic growth may initially lead to a deterioration of environmental quality. Only once a certain income threshold has been crossed do further increases in per capita income lead to gains in environmental quality. Increases in income from trade liberalisation may not necessarily translate into environmental improvements if the economic benefits are not shared with a reasonable degree of equitability among the population. So, international trade may become an important means of adaptation, but its effect is likely to be context-specific.[16]

The lack of clear WTO rules relating to clean energy and technology has increasingly complicated trade deals. Proposals to guarantee free trade in low carbon goods is one of the issues stuck in the Doha trade round, but may be pursued in the framework of bilateral trade agreements. So far, green trade has only been broached within bilateral accords and no general international agreement has been implemented, due to the complexity of reaching consensus among developed and developing countries. The WTO itself admits that 'The issue of climate change, per se, is not part of the WTO's on-going work programme and there are no WTO rules specific to climate change.'[17] WTO members have the right to adopt trade-related measures to protect the environment. But the concept of 'environmental protection' has not been taken to allow for the protection of green infant industries solely on the grounds of their being unable to compete with coal-fuelled counterparts in the short term.

Critics lambast the WTO for preventing preferential treatment in the trade of green technologies, which is necessary to speed up the transition to sustainable energy. This line of thinking sees international organisations and Western governments as being still too dominated by a free market ideology that is entirely unrealistic in presuming that mitigation and adaptation measures will be implemented by an invisible hand. Many see the current framework as inadequate for dealing with measures that promote the domestic production of green energy, as its definition of 'environmental protection' is too narrow.

Economic security is also related to investments in geo-engineering. Some fear that European governments are not investing enough in geo-engineering cooperation with international partners. Others believe they may be over-estimating the benefits of hoped-for technological solutions, that critics say simply divert attention from the underlying causes of global warming. There is real concern that there is a lack of international rules governing this area and that tensions may rise between states over the question of who controls the deployment of any massive geo-engineering projects that have an impact across multiple borders. Indeed, a nightmare scenario often conjured up is of states using their control over geo-engineering 'solutions' expressly to increase their power over other countries. China's national Weather Modification Centre has over 50,000 employees, and is seen as being ahead of any other actor in planning such interventions.

Interdependence as security?

European policies on these key questions have evolved notably. The geo-economics of climate change have become a more palpable policy priority. At the same time, however, positions remain ambivalent and divergent. Policy-makers recognise the read across to security dynamics but acknowledge that the linkages between the economics and politics of climate change are made in no more than cursory fashion. Two opposing trends co-exist within European policies. The first is a commitment to harness free markets as a means of assisting the development and spread of renewables. The implied logic of this policy dynamic is that a strategic model of cooperation and interdependence will be of most benefit to European governments in their attempt to manage the shift to low carbon.

Much of the European Commission's rhetoric is favourable to free market policies. It claims that:

> The EU is always looking for ways to reinforce its climate change goals with trade policies. Trade is how we will ensure the rapid spread of green goods,

> services and technologies around the world … Investment and trade can carry experience and skills and climate-friendly and low-carbon technologies throughout the global economy.[18]

The EU proposed an Environmental Goods and Services Agreement as part of the WTO Doha Round of trade talks. This aimed to liberalise trade in at least 43 key climate-friendly technologies identified by the World Bank. In 2007, the EU and the US attempted to implement a multilateral green trade regime, which was blocked by developing countries such as Brazil and Pakistan. Officials speak in private of the effort the EU has made in pressing for a new list of 'green goods' within the WTO to be opened up for liberalised exchange.

Basic EU guidelines stipulate that trade agreements must not undermine environmental goals; moreover, through the Generalised System of Preferences conditionality is exerted on partners' climate policy record and new accords depend on an enhanced set of Sustainability Impact Assessments.[19] Several member state governments have pressed for renewed effort to unblock 'green trade' in the Doha round, while also advocating that the EU move towards separate sectoral free trade on renewables. In December 2012 the EU finalised its first so-called 'Green Free Trade Agreement', with Singapore; this provides special trade rules for and lifts barriers in a range of green technologies. Including green clauses in bilateral FTAs has been seen as a means of circumventing paralysis within the WTO. In 2013 the EU looked to transfer the green trade provisions of the trade agreement with Singapore into other accords.

Some diplomats claim that the freeing up of trade in renewables is the most geo-strategically vital part of the climate security agenda. They have argued that the WTO needs to be brought into play to guard against the kind of export bans on food to which some governments have resorted since the droughts in 2009 and 2010. The German government's *Defence Policy Guidelines 2011* paper states that:

> Free trade routes and a secure supply of raw materials are crucial for the future of Germany and Europe. Restricted access can trigger conflicts. Disruptions of transport routes and the flow of raw materials and commodities, for example by piracy or the sabotage of air transport, pose a threat to security and prosperity. This is why transport and energy security and related issues will play an increasingly important role for our security.[20]

Germany has been the member state most strategic and most successful in developing a geo-economic policy to maximise the benefits from renewables exports. Germany has not only become the most successful exporter of renewables but also used its 'first mover' advantages to design commercially oriented regulations for the sector in the international trading system.[21]

The Dutch government supports the Institute for Sustainable Trade to keep the integrity of supply chains intact in states affected by climate instability. The UK's climate security envoy regularly opined that one of the most feared climate security risks is that the integrity of the global trading system will be compromised as states hunker down into a protectionist survival mode. The UK's 2011 Foresight report laid more stress on the geo-economic dimensions of climate change than any other previous official document. Climate change would be likely to disrupt trade routes and critical commercial infrastructure. The flooding of major Asian ports, for example, would seriously compromise the kind of global chains on which European economies are now highly dependent. Climate-related losses for insurance companies would hit the City especially hard, given its uniquely high exposure to global financial linkages. The short-termism of European financial institutions that was revealed by the post-2008 financial crisis could cost even more dearly in light of climate-induced disasters.[23] The FCO's climate unit has brought in trade expertise to address this under-focused issue. The July 2013 UK National Adaptation programme introduced a new and coordinated programme of investment, identifying 'adaptation services' as a niche area and fast-growing sector globally for UK companies.[24]

The EU has supported the US in launching an inquiry in the WTO against China in protest against its subsidies for domestic green technology producers. In October 2010, the EU joined Japan in protests against Ontario's Green Energy Act, a scheme that guarantees prices for renewable energy as long as it is generated with Canadian-made equipment. In 2011 Japan took this scheme to the WTO, claiming that it gave an unfair advantage to domestic manufacturers; the EU again protested in support. Officials talk of the common security and defence policy (CSDP) Atalanta anti-piracy mission off the Horn of Africa as being a climate-related geo-economic mission. The EU's growing engagement with concerns over Asian maritime security falls into the same category. More assertive action is driven in part by the desire to keep trade routes open.

Autarchy as security?

The second and contrasting trend in European policies is a reflex seeking to exempt renewables and green industry from market rules. The strategic logic here reflects the routinely made prediction that climate change will usher in a gradual retraction from global free markets. There is indeed evidence that at least some EU responses run contrary to market interdependence in an attempt to deal with the geopolitical uncertainties of climate change. While the EU's rhetoric is that of positive-sum market interdependence, at least some European policies betray a more mercantile outlook.

One leading team of experts detects signs of a more protectionist stance towards green trade.[25] In general there is little evidence that climate change is in fact driving EU external economic relations in a pro-liberal direction. It does not appear to have had any impact on the Common Agricultural Policy (CAP) reform – the latter clearly worsens scarcity in developing state and so feeds instability and conflict as climate impacts begin to hit. While the EU has not been the guiltiest party in Doha's failure, it has not shown flexibility in attempting new offers or sought to get the issue of renewable trade onto the agenda in a balanced and positive fashion. The EU has done little to ensure WTO coverage of renewables investment protection provisions. Crucially, member states reject any kind of commitment at the EU level that would oblige them to modify such economic policies in accordance with climate security indicators. DG Climate officials certainly admit that there has been no coordination between different parts of the EU decision-making machinery on the relationship between trade and climate security. Indeed, a general questioning of free market approaches has been increasingly evident.

Some parts of the EU policy-making machinery, and in particular those engaged in development work, question the value of open markets and view private investment in some cases as stirring up climate-related instability. They see international capital as supporting renewables initiatives that threaten to aggravate local social tensions. From a development point of view, officials insist that the imperative is to restrict big companies from buying up large tracts of land in poor countries to grow their own food in an effort to hoard resources as a preventative measure against climate change. These policy actors have increasingly sought to pull EU policies away from a free-market approach due to a generic concern over climate-related instability.

One highly respected expert closely monitoring trade issues concurs that by 2012 the trend was even more strongly towards green protectionism.[26] The EU has toughened up its Eco-Design Directive, with more stringent requirements on standards: some see this as laudable, others as discrimination against non-EU producers. A fierce debate persists between different actors within the Union over biofuels, with sharply differing views on the question of whether widening access for biofuels is environmentally ethical or strategically self-interested. The EU has increasingly used the 'indirect costs' process to impede ethanol imports from Brazil. The Renewable Energy Directive contains provisions on biofuels that some member states berate as acting in a protectionist fashion. The Commission has pushed to keep non-EU biofuels out of Europe through regulations to do with the biodiversity of where they are grown and the need for them to demonstrate at least a 35 per cent reduction in emissions – that is, justifying restrictions on environmental grounds as concerns over the 'indirect' impact of biofuels increase. However, observers expect these rules to be challenged in the WTO; the exemptions permitted where protectionism can be shown to benefit the environment may not apply to such a measure that merely slows the uptake of biofuels in Europe. Brazilian businesses complain that an EU tariff of 19 cents per litre of ethanol has prevented Brazil making a contribution to 20/20/20 targets.

France has focused increasingly on the export of nuclear technology, which other member states judge to be a security risk.[27] One of the most high-profile debates was over President Sarkozy's proposal for a 'green tariff'. After advocating such a measure in France, the president took his suggestion to Brussels, recognising that new protectionist measures were only feasible and only made sense at the Union level. After several years of debating such measures, there is still no clear, single EU line on carbon tariffs. Some European officials argue stridently that such trade-controlling measures must be part of the EU's pressure on third countries to meet emissions targets. Others express concern that they risk distorting investments and will serve as disingenuous pretext for a general return to protectionism. Rules on subsidies and discriminatory taxes have been pursued in an ad hoc free-for-all. The EU has imposed tariffs on some low carbon fuels from particular suppliers, like biofuels from the US, because they benefit from subsidies from their home governments, while member states make a more general case *in favour* of green subsidies.

While the G20 agenda formally includes the issue of elaborating rules on state subsidies for low carbon development, energy companies routinely berate the EU for doing little to restrict the scope of domestic subsidies in renewables as well as the oil and gas sector. The UK Foresight report points to some of the balances required: if 'green tariffs' are stringently outlawed, many states may argue they have less scope for domestic mitigation efforts.[28] The fact that the financial crisis has demonstrated in general terms the reality of market failure, this has spilled over to prompt urgency in climate security too. This points towards a finer balance between state intervention (to correct market mis-signals and provide capital) with market incentives (to get nations competing over renewables development).

The Commission has recently relaxed state aid rules to facilitate public investment in carbon capture and storage. Even the British government has moved towards more state intervention; significantly such a shift has been supported by all three main UK political parties. Critics have argued that wind farms have been massively over-subsidised, drawing resources artificially away from other areas of potential competitiveness gains, distorting global market dynamics with little apparent gain to emission reductions.[29] Access for renewables is now to be overseen by regulators, not the market. Germany and Spain in particular have favoured the route of feed-in tariffs as a preferential subsidy to renewables. These states also blocked the more market-based 'green certificates trading' system originally proposed by the Commission.

All such interventions serve to complicate third country access to European markets and in essence provide an export subsidy to help European firms conquer new global markets. Some observers observe the creep of a 'European champion' industrial strategy returning in a protectionist fashion that undermines market dynamism and competition. Free market advocates criticised the dilution of the Commission's unbundling proposals in 2009 as a huge environmental setback: breaking apart vertically integrated conglomerates would have made it easier for renewable sources to gain access to European grids. At the end of 2011, the International Energy Agency (IEA) stated that financial support to renewables was unacceptably high in nine EU member states and was a sign of new green mercantilism.[30] Confronted with criticism that mega-projects represent a spectacularly bad waste of taxpayers' money, EU officials have increasingly defended subsidies for renewables as cheaper in the long run than the extra costs to fuel bills as oil process inevitable rise.

Economists have calculated that the huge amount of subsidies going into green industry development needs to be better focused. They argue that subsidies are likely to be beneficial only where countries already possess some skills, existing expertise and infrastructure. They can help deepen a competitive advantage but not create it from nothing.[31] EU business representatives warn that the scale of European governments' domestic subsidies may contribute to a more general unravelling of at least the spirit of open global markets. And the focus on such large-scale subsidies also undermines prospects for the international extension of a carbon market, which many such companies see as more likely to provide a harmoniously governed system for tackling climate change than a zero-sum subsidies race. Criticisms are voiced that the extent of European subsidies now weakens the market mechanisms of the Emmissions Trading Scheme (ETS) and undercuts the EU's credibility when it seeks to encourage non-EU states to buy into the scheme's ostensible disciplines.

The formal and much repeated EU line is that liberal external energy policies flow naturally outwards from liberalisation of the EU's own internal market. However, this internal liberalisation appears to have stalled. While European leaders have committed to completing the EU energy market by 2014, implementation of the third package of liberalisation agreed in 2009 is not on track. The EU energy sector has actually become more, not less, concentrated, with increasing market shares in the hands of a small number of vertically integrated large conglomerates. In Germany the government has even over-turned rulings of its own anti-cartel office in order to enable mergers. Experts argue that the dilution of the Commission's unbundling proposals, combined with the similar withdrawal of the 'reciprocity clause', reflected a broader waning of the market dynamic in the EU's external relations too.[32]

Control or share?

These tensions – between liberalism and autarchy – express themselves in more specific concerns over European strategies for maximising shares of global renewable markets. This commercial question has a security read-over: is security best advanced by global partnerships to spread renewable technology as widely as possible or more narrowly by Europeans capturing the highest share possible of new renewables markets? Much is said about the shift to clean energy representing an opportunity in economic terms. Many policy statements and documents urge intensified effort to establish market control over low carbon technologies. The question arises: what is the relationship between such commercial opportunities and security imperatives? Are European governments guilty of seeking to protect market share and profits to the detriment of overarching geopolitical interest in disseminating new technologies? Or is the commercial momentum behind European policies contributing positively to the broader agenda of climate security?

As already indicated in previous chapters, on many measures, the EU has established a healthy position in the international development of low carbon. Up to 2011, the global top ten ranking of states' climate-related innovations contained four EU states – Germany in second place, France in fifth, the UK in sixth and Italy in tenth.[33] In 2010, 13 of the top 15 research companies in clean technology originated from the EU.[34] Spanish energy conglomerate Iberdrola is the second largest player in the US wind generation market. Denmark is the world leader in world turbine generation. The Desertec project is set to become one of the world's largest renewables initiative. *The Energy Roadmap 2050* states that decarbonisation should be a competitive boon for the EU as an 'early mover' in the global market for renewables.[35] The Foresight document reports a high level of confidence among the private sector and government that the UK is well positioned to gain from low carbon opportunities. This is an area in which European companies can once again compete, unlike the mass production manufacturing in which they have lost shares to rising powers. The UK financial sector is set to become a leading player in carbon trading, with potentially sizeable benefits for London.[36]

At the same time, it is widely acknowledged that Europe has more recently lost ground in green technology. China and the US have begun to steal a lead. Investors now rank European states' attractiveness as locations for renewable projects far lower in relative terms than a decade ago. Lord Nicholas Stern has warned that it is necessary for the EU to increase to a 30 per cent reduction in emissions as a self-interested investment in recouping declining shares of the green technology market from China, Taiwan and South Korea.[37] China exhibits increasing evidence of leapfrog growth, missing out some stages of economic development the West passed through. It is now one of the biggest producers of hydrogen cell cars; a leader in desalination technology; and has the world's biggest wind farm. Over half the solar panels in use in the EU now originate from China.[38] Some of the most notable green commitments are unfolding outside Europe. Abu Dhabi is building the world's first zero-carbon city, for example.

DG Energy has recognised that the Strategic Energy Technology Plan must be vastly upgraded.[39] In early 2011, energy commissioner Gunther Öttinger worried that 'At present, research and development are chronically underfunded in the EU. We risk missing the boat if there is a boom in the markets for new energy technologies.'[40] China's crisis-busting stimulus package allocated a higher share of resources to low carbon projects than did any EU member state's rescue package. Indeed, European leaders have explicitly recognised that China's new five-year plan overtakes the EU in terms of investment commitments in green technology.[41] In Spain renewables jobs have halved in the recession. The UK's flagship carbon capture and storage (CCS) project has been continually delayed.

So, how is the EU dealing with these mixed trends in its international geo-economic standing in renewables? There are some signs of the EU assuming an admirably pro-active role in promoting international cooperation in and sharing of low carbon technology. The EU is now setting up an international forum to cooperate on low carbon technologies, to include a large number of international partners, based, it says, on 'win–win reciprocity'.[42] An EU–GCC Clean Energy Network has been set up to share research in renewables. Euratom is collaborating with the US, Japan, China, India, Russia and Korea on a fusion generator; some predict that hydrogen fusion could become the key method of sustainable energy generation in the long term.[43]

The September 2011 Commission strategy on external energy policy highlighted new commitments in support of renewable energy sources beyond the EU's borders. These included a range of new international partnership agreements that included cooperation on renewables as their 'primary' aim; an initiative to encourage other forums such as the G20 to prioritise global rules on renewables development; reciprocity in access to renewables research programmes; an extension of the Energy Charter treaty's mandate to include rules on the renewables sector; a widening of the EU Energy Initiative on access to energy in Africa for the first time to include assistance on renewables; a strategic group for International Energy Cooperation made up of member states and Commission representatives; and a database of member states' energy projects in third countries.[44] A central aim was to combine renewables cooperation and broader economic support for growth, elements so far divorced from each other within European policies.[45]

In a sporadic sense, issues related to renewables are increasingly creeping into the broader panoply of European external relations. As a possible harbinger of future alliance-building priorities, Areva is spending 3 billion to increase uranium supplies from Canada, Kazakhstan, Niger and Namibia. The Scottish administration has turned to Middle East sovereign wealth funds for renewables investment: Masdar, a renewable energy company based in Abu Dhabi, plans to spend millions in clean energy projects in Scotland after signing an agreement with the Scottish government. The British and German foreign ministers launched a joint initiative in 2011 to encourage Russia to adopt firmer plans on energy efficiency. The EU's *Energy Roadmap 2050* called for a partnership with Russia and Ukraine especially on biomass.[46] Dutch firm Gasunie has signed a deal with Gazprom to import biomass energy.

Other aspects of policy point to a more defensive form of industrial mercantilism, however. Critics charge that European governments seem not to recognise that the strategic need to share new technologies is more important than the profits of a small number of private companies.[47] The EU has backed water privatisation schemes in developing states, with French and German companies in particular being successful in winning control of water resources. This has moved the Union away from support for the notion that access to water is a fundamental and top-priority human right, and towards a more geo-economic notion of resource control. One of the first petitions launched under the EU's new 2013 European Citizen's Initiative aimed to roll back this trend. In similar vein, most member states have rejected the Commission's push to limit biofuels on food security grounds, as governments see these as an easy way to meet more immediate emissions targets. Only the UK, Denmark, the Netherlands and Belgium have supported the Commission's stricter line on biofuels, for example seeking to report on how these add, rather than save, further emissions. Most states have launched strategies to gain more geo-economic control over biofuel sources.[48] The Commission's March 2013 green paper strikes a more geo-economic note in cautioning that the EU should no longer allow instruments like the Clean Development Mechanism to have European funds subsidise renewable competitors in rising powers.[49]

As China has increasingly insisted on green technology transfer in its bilateral trade and investment deals with European countries so the latter have sought tougher reciprocity. A 2011 EU trade initiative on reciprocity enshrined this shift in position. China's move into Greenland has been one factor in a rare earths dispute with the EU. While the EU and China are cooperating much more together on joint green technology projects and China is developing a cap-and-trade scheme modelled on the ETS, the EU and its member states have opened a number of cases against China's domestic support for various renewable projects. One former European climate minister argues that the EU must move much further in this direction, matching China in terms of harder-edged mercantilistic self-interest in green technology strategies. One expert points to the July 2013 agreement between China and the EU on 'green growth cooperation' as one of the major steps towards such a geo-economic approach. With the EU struggling to advance global multilateral rules to the environment, cooperation with rising powers on green trade and investment has increasingly been seen as the way forward, and yet such agreements include an uneasy mix of logics: 'The political weight of the EU will depend on whether this process of green modernisation will lead to closer cooperation on environmental problems or to intensified global competition on cutting edge technologies.'[50]

Officials advocate much more formal and high level political backing to sell renewable technology developed in Europe to other consumer countries around the world; that is, a more mercantile than developmental approach to climate funding. With such support, German commercial diplomacy has been strikingly successful, for example, in selling advanced solar panels to the Gulf, capable of working in extremely high temperatures. The 2011 external energy security strategy promises to increase political pressure behind the EU's Market Access Strategy to facilitate European renewable investments in emerging markets, and to focus on ensuring that standards in this sector reflect existing EU norms.[51] BusinessEurope has pressed for a more assertive geo-economic EU push to 'level the playing field', while rising powers refuse to sign up to post-Kyoto binding targets and while they see the EU's 'leadership by example' as failing.[52]

The mercantile line is evident in EU positions on intellectual property rules in relation to low carbon technologies. DG Trade has retained a firm line on this question. Its officials insist that green technology is not like medicines, where the EU has agreed to compulsory licensing agreements to allow low-cost production in developing states. They maintain that intellectual property rules are not the main obstacle to the broader spread of green technology; the problem lies in the limited uptake of green processes that have been around for many years and on which patents have already lapsed. Unlike medicines, research input is only a small share of production costs for renewables. A new renewable technology is, they argue, not comparable to a medicine where one drug can wipe out one disease. The green technology market is much more diverse and the availability of one new innovation will at best only have a small impact on the overall problem. In addition, it is impossible to define precisely what is and what is not a 'green' innovation. The line is blurred in the case of new production methods that have an energy savings component without being a renewable source per se. Policy-makers admit that they lean heavily on a consulting report which argued that intellectual property rights (IPR) have not restricted the spread of green technology; that there is sufficient competition in low carbon sectors to have pushed prices down, with no evidence that IPR have created monopolistic market positions; and that most green technologies are not covered by IPR.[53] The firm line is being maintained despite this complicating free trade talks.

Officials insist that developing states, through the Group of 77 (G77), use the debate over green IPR disingenuously as a bargaining chip to win more funding concessions from Europe. The incipient change is with China; China has been the toughest of all developing states against the EU on green IPR but as it is now becoming a source of green inventions it is shifting to a different position and seeing some benefit in firm IPR protection. Still, DG Trade officials express concern that DG Climate or environmental ministers may soften the line on IPR in comprehensive environmental talks in order to get multi-sector deals. The DG Trade approach is to support 'operational measures' to help spread new technology without diluting IPR; for example, spreading information to developing states of all the renewables freely available. Businesses lobby to claim that they would pull out of the green sector without IPR.

There are concerns that European governments have been unduly tempted to use scarce resources to favour indigenous firms rather than helping more international projects. Tensions have arisen with several non-European states over the terms of cooperation over renewables. The frequent complaint from developing countries has been that the EU is engaged in a quick grab for large-scale renewable projects oriented towards exporting energy to European markets rather than in a genuine partnership to maximise renewables' potential for 'host' societies too. ActionAid worries that European governments are pumping funds into large-scale, export-oriented renewables projects that are likely merely to worsen local conflict dynamics. This is especially the case with biofuels investments and their associated 'land grabs'. In an effort to assuage such discontent, the EU agreed that new projects under the rubric of the Union for the Mediterranean must ensure that 75 per cent of new renewable energy stays within North Africa. But the largest project in this area is Desertec, which remains outside the scope of such state-managed initiatives. Desertec is budgeted at nearly $600 billion and is driven by German companies who say it will meet 15 per cent of Europe's electricity needs by 2050. This project is seen in a particularly negative light in North Africa as cables will take the energy out into EU markets and not supply local demand; Desertec is often held up as symptomatic of an incipient eco-colonialism that could place severe new strains on relations between developed and developing countries. As one Moroccan insider puts it: Desertec is taking place 'in a vacuum', disconnected from the country's fragile domestic political context. Indeed, Morocco's semi-authoritarian regime insisted on sluicing off a share of the project into its own coffers.

Organisations like Counter-Balance admonish the EU for having been overly seduced by high-visibility geo-engineering projects for commercial self-interest. For critics, the solutions these promise are illusory. Indeed, relying heavily on a search for all-conquering technological breakthroughs could create more problems than it solves, to the extent that such a focus diverts policy-makers' attention from getting to the core drivers of climate insecurity. From this perspective, European governments stand accused of colluding too tightly with non-EU regimes on techno-fixes none of which exhibits convincingly proven potential, rather than targeting the more deep-seated governance pathologies that weave the most menacing links between climate change and geo-strategic tension.

Uncertainty and imprecision persist in this area of policy. In 2012, problems with a number of prospective geo-engineering projects revealed the absence of clear international rules to govern such initiatives. In response, the absence of clear EU positions was evident: member state governments expressed concerns over 'rogue research' while also funding their own explorations into geo-engineering solutions. A UK-funded initiative on aerosol injection was aborted in April 2012 due to uncertainty over international rules and intellectual property implications.[54] There has been no identifiable effort to develop cooperative rules at the international level for managing geo-engineering. Incipient explorations have taken place on an ad hoc basis, by private companies or on a national basis, often stirring up concerns across borders over the impact of such creeping unilateralism. DG Trade officials highlight how far they have led policy in this area with no input from the External Action Service (EAS) on how such mercantilism might impact the security agenda. Indeed, experts see governments as still lacking well-thought-through public policy strategies to cover geo-engineering and other alternative climate technologies. They berate Western powers for failing to knit together strategies for the dissemination of new technologies which combine interventions at the macro-multilateral level with local-level, micro-engagements at the community level; and accuse them of failing in this overdue multi-scalar approach precisely because they do not see the debates over different technological options fully as a matter of national security.[55]

Conclusion

In line with its deeper policy coordination in economic than hard security domains, the EU has taken the geo-economic dimensions of climate security increasingly seriously. This trend in policy suggests a broad and comprehensive conception of the climate security agenda. The seriousness with which questions of resource scarcity have been addressed militates against claims that the EU has an unduly narrow, militarised perspective on an essential socio-economic challenge. The EU's long-standing insistence that the economic and political dimensions of security cannot be separated finds an echo in the way that it has sought to contribute a distinctive added value to the climate security agenda. In this sphere, a marked continuity is evident between the EU's broader external relations identity and its extension into the field of climate security. The latter has been conceptualised in a way that provides a fillip to the EU's existing security norms, rather than contradicting or emasculating these.

Notwithstanding this, the policy record reveals that the EU still has a long road to travel before it can be said fully to have defined a set of geo-economic strategies that coherently and clearly address the security implications of climate change. The consensus is strong on the central importance of climate change's geo-economic impact, but less robust on what the appropriate way is to respond. European policies currently pull in contrasting directions: some responses have sought to deepen free trade commitments in order to smooth the flow of renewables; others betray governments' desire to keep a tighter control over trade and the harnessing of new technologies. Policy-makers suggest this reflects a necessary balance between market and state, between interdependence and protective national (or EU) autonomy. However, due to uncertainty over which leg of policy merits priority, the EU has moved decisively on neither front; climate concerns have not yet proven a major factor in explaining moves either for or against free trade. Moreover, there is a significant difference between governments backing their companies to maximise economic gains from renewables technology, on the one hand, and their prioritising a fully comprehensive understanding of economic security, on the other hand. On occasion, European governments come close to conflating these two strands of policy – that is, of reducing economic security to commercial advancement in green technology. In short, while European efforts to address the geo-economic dimensions of climate security have evolved notably, the policy mix remains ad hoc and often inchoate, with different aspects of incipient responses sometimes seeming to cancel each other out.

Notes

1 B. Lee, F. Preston, J. Kooroshy, R. Bailey and G. Lahn, *Resources Futures*, London, Chatham House, 2013, pp. 5–9.
2 J. Martin, *The Meaning of the 21st Century*, London, Eden Project Books, 2006, p. 303.
3 T. Jackson, *Prosperity without Growth*, London, Earthscan, 2009.
4 D. Helm, 'Green growth: Opportunities, challenges and costs', in L. Tsoukalis and J. Emmanouilidis (eds), *The Delphic Oracle on Europe: Is There a Future for the European Union?* Oxford, Oxford University Press, 2011.
5 S. Dalby, 'Climate change: New dimensions of environmental security', *RUSI Journal*, 158/3, 2013, 34–43.
6 C. Paskal, *Global Warring*, Basingstoke, Palgrave Macmillan, 2010, p. 20 and p. 95.
7 S. King, *Losing Control: The Emerging Threats to Western Prosperity*, Yale University Press, 2011.
8 Citi Bank, *Energy Darwinism: The Evolution of the Energy Industry*, London, Citi Bank, October 2013.
9 Martin, *The Meaning*, p. 302.
10 European Commission, *EU2020: Europe's Growth Strategy*, Brussels, European Commission, 2010.

11 UNEP, *Climate and Trade Policies in a Post-2012 World*, Geneva, UNEP, 2009.
12 European Commission, *A Stronger European Industry for Growth and Economic Recovery*, Brussels, COM (2012) 582.
13 C. Huhne, 'The geopolitics of climate change', speech at the Royal United Services Institute, 7 July 2011.
14 Defra-BIS, *Resource Security Action Plan: Making the Most of Valuable Materials*, London, The Stationery Office, 2012.
15 National Intelligence Council, *Global Trends 2030: Alternative Worlds*, Washington, NIC, 2012, p. 31.
16 WTO-UNEP, *Trade and Climate Change*, Geneva, WTO, 2009.
17 www.wto.org/english/tratop_e/envir_e/climate_intro_e.htm.
18 http://ec.europa.eu/trade/wider-agenda/environment/climate-change/.
19 www.ec.europa.eu/trade/wider-agenda/environment.
20 Federal Ministry of Defence of Germany, *Germany: Defence Policy Guidelines 2011*, Defence white paper, Federal Ministry of Defence of Germany, 2011.
21 M. Janicke, 'German climate change policy: Political and economic leadership', in R. Wurzel and J. Connelly (eds), *The European Union as a Leader in International Climate Change Politics*, London, Routledge, 2011.
23 Foresight, *International Dimensions of Climate Change*, London, UK Government Office for Science, 2011, p. 54.
24 Her Majesty's Government, *UK National Adaptation Programme: Making the Country Resilient to a Changing Climate*, London, The Stationery Office, p. 7.
25 Overseas Development Institute, *The Next Decade of EU Trade Policy*, London, ODI, July 2012.
26 F. Erikson, 'The rising wave of green protectionism: The EU and biofuels', ECIPE working paper 2, 2012.
27 J. Szarka, 'France's troubled bids to climate leadership', in Wurzel and Connelly, *The European Union as a Leader*, p. 117.
28 Foresight, *International Dimensions*, p. 56.
29 J. Constable, *The Green Mirage: Why a Low-carbon Economy may be Further off than we Think*, London, Civitas, 2011.
30 Platts *EU Energy*, Issue 271, 2 December 2011, p. 3.
31 M. Huberty and G. Zachmann, 'Green exports and the global product space: Prospects for EU industrial policy', Brugel working paper 7, 2011.
32 P. Eikeland, 'EU internal energy market policy: Achievements and hurdles', in V. Birchfield and J. Duffield (eds), *Toward a Common European Union Energy Policy*, Basingstoke, Palgrave, 2011, p. 26.
33 A. Bowen and J. Rydge, 'Climate change policy in the United Kingdom', Grantham Research Institute for Climate Change policy paper, August 2011, p. 17.
34 D. Hamilton, *Europe 2020: Competitive or Complacent?*, Washington, Centre for Transatlantic Relations, p. xvii.
35 European Commission, *Energy Roadmap 2050*, COM (2011) 885/2, p. 9.
36 Foresight, *International Dimensions*, p. 93.
37 N. Stern, 'The EU risks falling behind in the green century', *Financial Times*, 5 July 2011, p. 9.
38 Platts *EU Energy*, Issue 290, 18 November 2011, p. 9.
39 DG Energy, *Energy 2020*, Brussels, European Commission, 2011, p. 18.
40 New Europe, 'Our world in 2011', 7 January 2011.
41 H. van Rompuy, 'Challenges and priorities for EU energy policy', speech at Strategic Energy Forum, Brussels, 11 May 2001, European Council, PCE 0111/11.

42 Platts *EU Energy*, Issue 246, 3 December 2010, p. 9.
43 Martin, *The Meaning*, p. 163.
44 European Commission, *On Security of Supply and International Cooperation – The EU Energy Policy: Engaging with Partners beyond our Borders*, COM (2011) 539.
45 European Commission, *Results of the Public Consultation on the External Dimension of the EU Energy Policy*, SEC (2011) 1023.
46 European Commission, *Energy Roadmap 2050*, p. 11.
47 N. Mabey, 'Climate change and global governance', E3G memo, October 2009, www.e3g.org, p. 7.
48 *Euractiv*, 25 March 2013.
49 European Commission, *Green Paper: A 2030 Framework for Climate and Energy Policies*, COM (2013) 169, p. 10.
50 A. Ruser, 'Setbacks in climate change negotiations may signal dire straits for normative power Europe', *LSE Europe blogs*, 3 October 2013.
51 European Commission, *The EU's Energy Policy*.
52 BusinessEurope, 'Doha climate conference: Absence of level playing field requires EU climate and energy policy rethink', press release, 10 December 2012.
53 Copenhagen Economics, *Are IPR a Barrier to the Transfer of Climate Change Technology?* Copenhagen, Copenhagen Economics, January 2009.
54 International Institute for Strategic Studies, *Strategic Comments*, Volume 18, Comment 46, 12 December 2012.
55 M. Brown and B. Sovacool, *Climate Change and Global Energy Security: Technology and Policy Options*, Boston, MA, MIT Press, 2011.

9

CONCLUSIONS

This book's rationale has been to move debate beyond broad-brush predictions of how climate change will affect security towards an assessment of how security policies have begun to respond in practice to these challenges. Warnings that climate change must be conceived as a security challenge have been forthcoming for more than a decade. Over that decade, many governments and international organisations have rhetorically assented with this prognosis and committed themselves to developing effective strategies of climate security. In this book, I have tested the commitments of one such set of actors, the EU and its member states.

The book's opening chapters proposed a set of operational guidelines for assessing EU climate security policy. Two overarching questions were presented. First, how much priority has been attached to climate security policies? Second, have these policies tilted more towards cooperative–holistic or realist–defensive notions of security? Chapter two suggested that a comprehensive climate security strategy would be expected to embrace a range of challenges: to prepare for climate-related conflict resolution; to develop more structured diplomacy and incentives focused on climate-related global dynamics; to adapt economic and development policies specifically to temper climate-induced risks; to manage new patterns of international alliances; to move away from purely state-centred approaches to security; and to design more agile response mechanisms to deal with future uncertainty.

In line with these challenges, chapter three outlined the processes and identities that have defined the EU's role in foreign and security policy and related these to a set of alternative policy outcomes. A *realist or rivalry*-oriented approach would be manifest in additional investment in defensive military capability; increased tension with strategic partners over climate security differences; a more hard-headed form of geo-economics; tougher realpolitik against those states seen as problematic from a climate security perspective; resources dedicated to protecting national borders and repelling climate migrants; and a drift towards self-help independence. In contrast, the policy outcomes reflecting a *liberal–cooperative* approach would include enhanced efforts at multilateral partnership; resources invested in civilian conflict prevention and underlying governance reforms in developing states; increased development cooperation and a more rules-based set of international economic policies; a wider network of cooperative security partnerships; and a more participatory system of security governance. These two approaches are useful as heuristic schema; in practice, the challenge is to determine the mix and combination of realist and liberal dynamics within European security strategies.

In broad-brush terms, the evidence invites a mixed assessment of each of these operational questions. In terms of their formal strategic visions and planning, European governments have certainly taken climate security increasingly seriously. Far from under-reacting, they have increasingly defined climate security as a linchpin issue that conditions all other areas of geo-strategy. Very little climate security denial remains amongst European politicians and diplomats; the doubts over climate change's strategic impact that were outlined in chapter two resonate little among policy-makers. The EU's conceptual understanding of climate change's security ramifications has become stronger, more sophisticated and admirably nuanced. The issue has progressively seeped into the policy mindset across the various different ministries, departments and commission directorates whose responsibilities have some bearing on climate challenges.

However, a range of factors has weighed against the practical advance of climate security policies. The EU has found itself confronted with so many more immediately urgent challenges – from the economic crisis to the Arab Spring – that in practice the issue of climate security has slipped down its list of priorities. It seems destined to be an issue that is perennially just over the horizon, requiring attention … but never right now. The EU's well-known institutional complexities work to the disadvantage of the kind of inter-issue 'linkage' challenges of which climate security is a prime case. The aftermath of the Lisbon Treaty has, if anything, seen such complexities multiply; climate security has fallen between the wider 'cracks' between different lines of institutional responsibility. The attention paid to traditional understandings of hydrocarbon-oriented energy security still far exceeds the more nebulous preoccupations of climate security.

In general terms, the cooperative framework is formally more pre-eminent in incipient EU policies than the set of outcomes that would be expected of a realist–rivalry approach. European governments have made the judgement that climate insecurity renders deeper multilateral cooperation necessary. Indeed, this issue has to some extent acted as a counter-balance to the creep of more introverted–realist dynamics in other areas of EU security policy. Policy-makers claim, for example, that the significant increases in European climate financing offered to poorer countries reflect both a greater degree of commitment and the prevalence of a cooperative, development-oriented approach. Climate security is an issue on which the EU has made a genuine attempt to recover proactive leadership in advancing comprehensive and internationalist understandings of security at a global level.

The approach incrementally mapped out has been holistic in nature, consistent with the EU's traditional stress on civilian power instruments and its perceived need to legitimise foreign policy positions by reference to claims of justice. There are fears that militaries have disingenuously over-reacted in an attempt to use the climate security agenda as a means of reinforcing their own claim to resources and influence within government. While such concerns are not entirely unfounded, however, there is insufficient evidence so far to sustain the claim that EU climate security policies have become overly militarised. In the US the military lead on climate security is much more striking than in Europe. The problem with armed forces' engagement is rather that this has been limited to relatively narrow questions of 'greening' military operations; militaries have inched toward broader geopolitical deliberation, but so far in a more cautious manner. In the military sphere, policy outcomes do not – yet – substantiate the realist–rivalry framework.

Indeed, the EU's weakness may run in the opposite direction. Far from over-securitising environmental questions, the EU rather easily assumes that it has such a strong record in terms of its mainstream climate change policies – emissions targets, renewable development – that it automatically makes a strong contribution to climate security at a global level. The EU's internal policy initiatives and integration clearly do have a broader international spill-over. But the EU tends rather uncritically to assume that the extension of its own rules and templates axiomatically constitutes a security policy beyond its borders. This is an overly reductive and linear assumption, which also exaggerates the strength of core, internal EU climate change commitments. In many ways climate change remains under-securitised. Despite much apocalyptic rhetoric, policy initiatives have often come down, somewhat incongruously, to the EU proffering relatively mundane matters of energy efficiency. European policy-makers are fond of exhorting that changing light bulbs, funding loft insulation or getting militaries to re-use equipment are the best contributions to climate-sensitive foreign policy. While these elements may be an important part of a comprehensive response, arguably the EU still exaggerates how far they really contribute to the high politics of a genuine security policy.

Through a realist lens all this would denote a 'failed securitisation'. However, the picture is better described as a 'securitisation-lite', whose shortcomings are of a more subtle nature: namely, that, paradoxically, the policy outcomes associated with the liberal–cooperative model have themselves been limited by their lack of truly security-related logic. The EU has progressed further in addressing human security vulnerabilities through its adaptation-orientation funding initiatives than it has in tackling the broader international relations of climate security. This is not to advocate any kind of hard-power, realpolitik approach. Quite the contrary; it is to suggest that the EU's own liberal–cooperative aspirations still require greater political security commitment and tenor if they are to be fully realised.

Notwithstanding the tilt towards positive-sum solutions, European governments are hedging between deeper international cooperation and self-reliance. Policy outcomes certainly do not yet appear wholesale to reflect arguments that the only way to guarantee security in a world ravaged by extensive warming is through the 'lifeboat' solution. European policy-makers formally reject this argument, arguing that it would be neither feasible nor desirable to pull up the barricades and disconnect from the international system. There has certainly been no massive redirecting of military resources towards 'defending the homeland' against a climate-scarred world. Yet elements of EU policies do suggest that governments are at least leaving open the option for some greater degree of self-reliance and the need for a more rivalry-based geopolitics. While to some extent this is understandable and may even prove wise, it means that policy outcomes often exhibit an uneasy commingling of the rivalry and cooperative frameworks.

This is seen most notably in the nature of migration policies. It is true that climate change is not the major factor driving a tightening-up of EU migration rules. But it does feed into and magnify existing pressures against open borders. The EU has not been willing to recognise any putative legal category of environmental refugees. There is also a sceptical view of European climate financing: many diplomats themselves fear that this remains too limited to serve strategic ends and is increasingly directed at containment-oriented adaptation projects that are unlikely to help lay the foundations for long-term security. In both these areas, policy outcomes cannot powerfully be explained by a form of sufficiently broad climate security deliberation, either in a rivalry or cooperation mode.

European policy-makers are acutely aware of the dangers that climate change presents to conflict dynamics in fragile states and societies. Climate change is judged to augment the importance of 'doing' conflict prevention and resolution. Yet, in practice the climate security agenda has not yet ensured a quantitatively greater EU commitment to conflict prevention. Nor has it had any major qualitative impact on the way in which the EU approaches conflict prevention. While there is little evidence that climate security concerns have caused any significant diversion of governance-oriented funding into, for example, defensive migration controls, neither has it given a spur to policies aimed at underlying causal dynamics. Where resources and diplomacy have been drawn away from liberal–institutionalist to containment-based strategies, climate security has rarely been the reason for such a drift. Indeed, while policy-makers routinely opine that climate security over-shadows the importance of all other aspects of security, there is no sign that it has been a pre-eminent factor in determining the way in which external EU funding is allocated and spent.

One of the few areas that has advanced some climate-conflict specificity is that of the EU's 'access to energy' initiatives. These exhibit much admirable foresight and have been useful at a micro-level in safeguarding local communities' access to energy supplies. But they remain limited in scope and are often undercut by other dimensions of EU policies oriented more to 'extracting' such energy resources for European use. This reflects a fundamental and unresolved tension at the heart of European policies. Is security primarily a matter of bringing scarce energy resources into Europe? Or is it a question of ensuring that the latter are distributed, both internationally and within locally fragile contexts, in the kind of equitable fashion that reduces the likelihood of conflict? So far, policy outcomes suggest that the EU has sought to straddle these two approaches without a clear prioritisation or acknowledgement of the trade-offs involved.

The EU's understanding of the geo-economic dimensions of climate security is also inconstant. Liberal arguments have both a cause and effect side: free trade is needed to generate and spread low carbon technology, and thus guarantee climate security; at the same time, a major fear is that one of the main strategic effects of climate change is that the liberal economic order will collapse, requiring a far stiffer defence of such an order in the short term. Rhetorically, the EU collectively and most European governments individually concur with this liberal perspective on the geo-economics of climate change. They have made some effort to embed liberal 'green economics' principles in organisations such as the World Trade Organisation (WTO). Indeed, there are some justifiable concerns that a faith in liberal economic globalism still needs to develop a more nuanced form to mould itself to climate-related imperatives.

Yet European governments have also tended increasingly towards more mercantilistic policies in practice. They have done relatively little in concrete terms to prioritise a defence of the liberal trading system against the supposed threat from climate change. Geo-economic security is increasingly seen in terms of control over low carbon technology, alongside the nominal commitment to global market inter-dependence. The evidence does not sustain the critical perspective that Western governments' alleged addiction to liberal economics is so unconditional that they are blind to security concerns. Moreover, the EU sees security in terms of exporting its own energy-market regulations; while this approach has been helpful in extending some climate-friendly standards, it has also complicated more agile climate security policies. As in other areas, the EU approach is best described as an imprecise hedging between liberal and realist approaches.

Finally, climate change has not yet ushered in a new model of EU security governance. European governments have reached out to companies and civil society organisations, and acknowledged that climate security requires a less state-centric template. Yet the interaction between governments, the private sector and non-governmental organisations (NGOs) on security issues remains relatively limited. Few policy initiatives have been launched that provide companies and civic organisations with formalised input into climate security deliberations. There remains a feeling that climate security, like climate change policy in general, requires more top-down processes to counter public and commercial narrow self-interest. Most companies and NGOs themselves remain focused on traditional, core environmental concerns and often express scepticism over the climate security agenda.

This mix of 'rivalry' and 'cooperation' dynamics is integrally related to the EU's *sui generis* security identity mapped out in chapter three. This mixed identity helps explain the policy outcomes that the book identifies; it is also itself reinforced by these policy outcomes. Moreover, the balance of progress and shortfall in evolving European policies points the way towards what is still needed for a more effective climate security strategy. The EU still needs to make the move from 'climate security' to a full-spectrum climate foreign policy. While the self-defined security community has begun to come to terms with the implications of global warming, efforts more broadly to mainstream climate within foreign policy as a whole need to be markedly intensified. The EU requires an assessment in the foreign policy sphere of what the Stern review did in the economic sphere in terms of unpacking the costs of not acting on climate change.

The EU must be right in rejecting rivalry-lifeboat scenarios, but still needs to articulate a climate security policy around more nuanced and cleverly crafted give-and-take strategic alliances, predicated on the premise that both rich and developing states need each other to make progress on this agenda. Many remain sceptical of the climate security agenda because they conflate securitisation with militarisation; but the fact that the latter should be limited does not mean that the former is not needed. More strongly embedded international regimes will be required in all areas of security and crisis management. And the EU must still move beyond climate security being debated merely in terms of Europe having to defend itself against instability that originates 'out there' as opposed to mitigating the way its own policies often magnify global threats. Security will not be ensured simply by spending modestly higher amounts on conflict prevention, adaptation or renewables-dissemination in developing states, however necessary and welcome such steps are. More profound adjustment is pending on the very essence of the way in which EU foreign and security policy interacts with and seeks to shape the global order. It is here that the opportunity lies for the EU to build upon advances made in the last decade to assume more far-sighted leadership on issues of global climate security.

INDEX

Made in the USA
Las Vegas, NV
19 September 2026

b96fd6f1-dd85-45a7-8b23-c0921c79864fR01